ENERGY AND
THE ENVIRONMENT

Biofuels

ENERGY AND THE ENVIRONMENT

Biofuels

JOHN TABAK, Ph.D.

Facts On File
An imprint of Infobase Publishing

 *To my parents, Claire and Leo Tabak*

Facts On File, Inc.
An imprint of Infobase Publishing
132 West 31st Street
New York NY 10001

Library of Congress Cataloging-in-Publication Data
Tabak, John.
  Biofuels / John Tabak.
    p. cm.—(Energy and the environment)
  Includes bibliographical references and index.
  ISBN-13: 978-0-8160-7082-4 (acid-free paper)
  ISBN-10: 0-8160-7082-2 (acid-free paper)
  1. Biomass energy. I. Title.
  TP339.T33 2009
  662'.88—dc22   2008024349

Facts On File books are available at special discounts when purchased in bulk quantities for businesses, associations, institutions, or sales promotions. Please call our Special Sales Department in New York at (212) 967-8800 or (800) 322-8755.

You can find Facts On File on the World Wide Web at http://www.factsonfile.com

Text design by Erik Lindstrom
Illustrations by Jeremy Eagle
Photo research by Elizabeth H. Oakes

Printed in the United States of America

Bang 10 9 8 7 6 5 4 3 2 1

This book is printed on acid-free paper.

 # Contents

Preface

Nations around the world already require staggering amounts of energy for use in the transportation, manufacturing, heating and cooling, and electricity sectors, and energy requirements continue to increase as more people adopt more energy-intensive lifestyles. Meeting this ever-growing demand in a way that minimizes environmental disruption is one of the central problems of the 21st century. Proposed solutions are complex and fraught with unintended consequences.

The six-volume Energy and the Environment set is intended to provide an accessible and comprehensive examination of the history, technology, economics, science, and environmental and social implications, including issues of environmental justice, associated with the acquisition of energy and the production of power. Each volume describes one or more sources of energy and the technology needed to convert it to useful working energy. Considerable emphasis is placed on the science on which the technology is based, the limitations of each technology, the environmental implications of its use, questions of availability and cost, and the way that government policies and energy markets interact. All of these issues are essential to understanding energy. Each volume also includes an interview with a prominent person in the field addressed. Interview topics range from the scientific to the highly personal and reveal additional and sometimes surprising facts and perspectives.

Nuclear Energy discusses the physics and technology of power production, reactor design, nuclear safety, the relationship between commercial nuclear power and nuclear proliferation, and attempts by the United States to resolve the problem of nuclear waste disposal. It concludes by contrasting the nuclear policies of Germany, the United States, and France. Harold Denton, former director of the Office of Nuclear Reactor Regulation at the U.S. Nuclear Regulatory Commission, is interviewed about the commercial nuclear industry in the United States.

Biofuels describes the main fuels and the methods by which they are produced as well as their uses in the transportation and electricity-production sectors. It also describes the implications of large-scale biofuel use on the environment and on the economy with special consideration given to its effects on the price of food. The small-scale use of biofuels—for example, biofuel use as a form of recycling—are described in some detail, and the volume concludes with a discussion of some of the effects that government policies have had on the development of biofuel markets. This volume contains an interview with economist Dr. Amani Elobeid, a widely respected expert on ethanol, food security, trade policy, and the international sugar markets. She shares her thoughts on ethanol markets and their effects on the price of food.

Coal and Oil describes the history of these sources of energy. The technology of coal and oil—that is, the mining of coal and the drilling for oil as well as the processing of coal and the refining of oil—are discussed in detail, as are the methods by which these primary energy sources are converted into useful working energy. Special attention is given to the environmental effects, both local and global, associated with their use and the relationships that have developed between governments and industries in the coal and oil sectors. The volume contains an interview with Charlene Marshall, member of the West Virginia House of Delegates and vice chair of the Select Committee on Mine Safety, about some of the personal costs of the nation's dependence on coal.

Natural Gas and Hydrogen describes the technology and scale of the infrastructure that have evolved to produce, transport, and consume natural gas. It emphasizes the business of natural gas production and the energy futures markets that have evolved as vehicles for both speculation and risk management. Hydrogen, a fuel that continues to attract a great deal of attention and research, is also described. The book focuses on possible advantages to the adoption of hydrogen as well as the barriers that have so far prevented large-scale fuel-switching. This volume contains an interview with Dr. Ray Boswell of the U.S. Department of Energy's National Energy Technology Laboratory about his work in identifying and characterizing methane hydrate reserves, certainly one of the most promising fields of energy research today.

Wind and Water describes conventional hydropower, now-conventional wind power, and newer technologies (with less certain futures) that are being introduced to harness the power of ocean currents, ocean waves, and the temperature difference between the upper and lower layers of the ocean. The strengths and limitations of each technology are discussed at some length, as are mathematical models that describe the maximum amount of energy that can be harnessed by such devices. This volume contains an interview with Dr. Stan Bull, former associate director for science and technology at the National Renewable Energy Laboratory, in which he shares his views about how scientific research is (or should be) managed, nurtured, and evaluated.

Solar and Geothermal Energy describes two of the least objectionable means by which electricity is generated today. In addition to describing the nature of solar and geothermal energy and the processes by which these sources of energy can be harnessed, it details how they are used in practice to supply electricity to the power markets. In particular, the reader is introduced to the difference between base load and peak power and some of the practical differences between harnessing an intermittent energy source (solar) and a source that can work virtually continuously (geothermal). Each section also contains a discussion of some of the ways that governmental policies have been used to encourage the growth of these sectors of the energy markets. The interview in this volume is with John Farison, director of Process Engineering for Calpine Corporation at the Geysers Geothermal Field, one of the world's largest and most productive geothermal facilities, about some of the challenges of running and maintaining output at the facility.

Energy and the Environment is an accessible and comprehensive introduction to the science, economics, technology, and environmental and societal consequences of large-scale energy production and consumption. Photographs, graphs, and line art accompany the text. While each volume stands alone, the set can also be used as a reference work in a multidisciplinary science curriculum.

Acknowledgments

The author gratefully acknowledges the help received from Jim Allen in the media relations department at the Tennessee Valley Authority and from John Irving of the Burlington Electric Department, for their assistance in researching some of the technologies employed by their respective companies. Thanks also to Frank Darmstadt, executive editor, Facts On File, for his patience and support; Elizabeth Oakes, for her fine photo research; and Dr. Amani Elobeid, for the generous way that she shared her many ideas and insights.

Introduction

Obtaining sufficient supplies of energy in an environmentally responsible way is one of the central problems of the 21st century. Enormous resources will be required to produce a solution, and every proposed solution will be controversial. There is no easy answer.

Of all the problems associated with energy production, none is more complex than those associated with the production of *biofuels*, nonfossil fuels derived from *biomass*, a term that includes plant matter, animal wastes, and municipal wastes. Obtaining enough biomass to meet demand will require that large tracts of forest be intensively managed, and it will also require that agricultural resources are diverted from the production of food and feed and toward the production of fuel. Biofuels entail their own special kind of environmental and economic disruption.

This volume describes the most common types of biofuels and biofuel technologies and seeks to identify both the advantages and disadvantages associated with their use. There are many types of biofuels, and they are used in many different ways. Most famously, biofuels are used in the transportation sector, usually blended with gasoline or diesel fuel, but they are also used to generate electricity and provide heat for residential and industrial use. Biofuel consumption may be as simple as throwing a log on a fire or as complex as burning a combustible gas derived from plant matter in a combined-cycle electricity-generating unit. This book seeks to capture some of that diversity.

The first chapter describes what biofuels are and then describes some of the history of humanity's use of wood. This brief history is complicated by the fact that there are more reasons to burn wood than to obtain heat, and more reasons to acquire wood than to burn it. These histories illustrate many of the same problems nations face today when they attempt to use biofuels. Some earlier peoples succeeded in using this most elemental fuel source responsibly, and some did not.

While there are many types of biofuels, all biofuels are produced to be burned. They can, therefore, be compared on the basis of their energy content as well as their cost, their availability, and the environmental consequences associated with their production and consumption. These ideas are introduced in chapters 2 and 3, but they are addressed in the other chapters as well. Without an appreciation of these ideas, it is difficult to understand both the advantages and limitations of biofuels.

In the United States, roughly 90 percent of all petroleum is used in the transportation sector, and most cars, trucks, trains, planes, and ships burn petroleum to generate their power. Petroleum consumption and transportation are almost synonymous. The biofuels ethanol and biodiesel are two of the very few alternatives to petroleum. Compared to the size of the petroleum market, the markets for ethanol and biodiesel are tiny, even though they have been in use for as long as gasoline and diesel fuel. (Some of the first automobile engines burned ethanol, and some of the first diesel engines burned vegetable oil.) Chapters 4 and 5 discuss ethanol and biodiesel, respectively.

Chapter 6 describes the ways that biofuels are currently used to generate electricity. Currently, very little electricity is generated through the consumption of biofuels. There are good reasons that biofuels make only a small contribution to this vital industry. This chapter describes the current state of the technology.

Biofuels can be solid, liquid, or gaseous. They are remarkably versatile, and yet there are severe constraints on their large-scale use. Chapter 7 describes some of the basic physical constraints on the large-scale use of biofuels. But even in situations where the maximum potential contribution of biofuels to the energy mix is small, there may still be compelling reasons for using them. Understanding why this is true is important if one is to fully appreciate the environmental advantages of using biofuels. Chapter 8 gives several examples of how biofuels have been used to solve environmental problems while simultaneously contributing to the production of energy.

In developed countries, the ethanol and biodiesel industries exist only as the result of determined government intervention in the transportation fuels market. Generous subsidies for ethanol and biodiesel production are common, as is the requirement that these fuels be mixed with their petroleum counterparts in order to guarantee a market for biofuel producers. This is true in the United States, in the European Union, and even in Brazil, the foremost pioneer in the creation of a modern biofuel transportation market. Chapter 9 discusses some of the measures that the United States government has undertaken to attempt to develop a biofuel market. Government efforts to create biofuels markets have not always been successful, but they have always been expensive. They demonstrate just how hard it is to displace even a small amount of gasoline or diesel with the corresponding biofuel.

The production and consumption of biofuels involve some of the most complex and interesting problems in the energy sector. The solutions to these problems will have important economic and environmental consequences, and they will be part of the increasingly animated public discussion about energy and the environment.

PART I

The Nature and History of Biofuels

First Fuels

Biofuels, nonfossil fuels produced from plant materials and animals wastes, were the first fuels, and they have been in use for a very long time. There is evidence that fire was used for cooking by modern humans living along the coast of South Africa 164,000 years ago. Even Neanderthals, the closest relatives to modern humans, apparently used fire, although the details of the way they used it remain unclear. Biofuels were probably the only source of energy under human control in distant times, and they continue to make an important contribution today.

The responsible and intensive use of biofuels depends on solving a number of technical and environmental problems associated with their use. These problems, some of which are, in retrospect, apparent even in the histories of the world's oldest societies, reveal a great deal about the nature of biofuels. This chapter describes some early efforts to use wood. For hundreds of millions of people alive

Biofuels were the first energy source to be brought under human control. They remain an important source of energy today. *(Einer Helland Berger)*

today, many of these early technologies and practices are familiar because they are still in daily use. More generally, the early history of wood as a fuel illustrates that large amounts of wood are needed in order to sustain even a modest population, and without intensive forest management, the production of this wood can result in environmental disasters. In order for biofuels to become important sources of energy, biomass, the material from which biofuels are manufactured, must be harvested on an astonishing scale, and the harvest must be used efficiently. Inefficiency will result in widespread environmental destruction.

This chapter begins with a description of what biofuels are, because biofuels are not one fuel source but many. The importance of

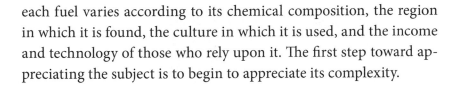

each fuel varies according to its chemical composition, the region in which it is found, the culture in which it is used, and the income and technology of those who rely upon it. The first step toward appreciating the subject is to begin to appreciate its complexity.

WHAT ARE BIOFUELS?

Biofuels are fuels that are obtained from organic materials. They include, but are not limited to, wood, straw, sugarcane residue, animal wastes, *landfill gas,* biodiesel and ethanol, municipal waste, *black liquor* (an energy-rich liquid produced by the *pulp* and paper industry), *switchgrass,* and a variety of synthetic gases derived from plant and animal matter. Biofuels are distinct from fossil fuels in that they are, in some sense, renewable. There are so many biofuels, and they are derived from so many sources, that their description requires a classification scheme. In fact, biofuels are commonly classified in several ways, three of which are described here. The first classification scheme uses three categories: (1) wood fuels, (2) agricultural fuels, and (3) urban waste-based fuels.

In addition to fuel wood, the category "wood fuels" includes charcoal, which is manufactured by partially combusting wood in a low-oxygen environment, and black liquor. Black liquor, an energy-rich liquid derived from wood, is burned by the companies that produce it, in part to obtain the energy needed to continue manufacturing operations and in part to recover noncombustible materials suspended within it. Wood can also be used as a *feedstock,* or raw material, to manufacture a combustible gas through a process called *gasification* or to produce ethanol, although neither the gasification process nor the wood-to-ethanol conversion process has reached the point where the technology is ready to be commercialized. There is, however, a great deal of interest in both processes. Gaseous and liquid fuels are often preferred to direct combustion of solid plant matter because they tend to burn more cleanly and can be used in ways that a solid fuel cannot. Wood fuels are, then,

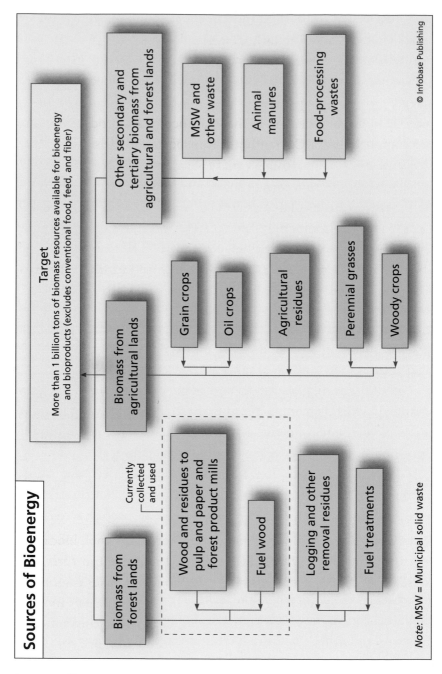

Sources of bioenergy (Source: U.S. Department of Energy Genome Programs)

Sources of Bioenergy

Target
More than 1 billion tons of biomass resources available for bioenergy and bioproducts (excludes conventional food, feed, and fiber)

Other secondary and tertiary biomass from agricultural and forest lands

- MSW and other waste
- Animal manures
- Food-processing wastes

Biomass from agricultural lands

- Grain crops
- Oil crops
- Agricultural residues
- Perennial grasses
- Woody crops

Biomass from forest lands

Currently collected and used

- Wood and residues to pulp and paper and forest product mills
- Fuel wood
- Logging and other removal residues
- Fuel treatments

Note: MSW = Municipal solid waste

© Infobase Publishing

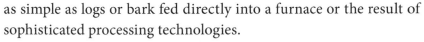

as simple as logs or bark fed directly into a furnace or the result of sophisticated processing technologies.

Agricultural fuels generally fall into one of two categories: specially grown crops and agricultural waste. In the specially grown category are fuel crops such as switchgrass, which can be burned directly to produce heat or, in theory, processed to produce ethanol or a combustible gas. Currently, the transportation fuels ethanol and biodiesel are obtained from a variety of agricultural crops—ethanol is often produced from corn or sugarcane, and biodiesel is produced from many kinds of plants as well as grease from restaurants and food-processing companies. In the agricultural waste category are animal wastes, corn *stover,* which is the part of the plant left behind after the corn kernels have been harvested, and bagasse, the part of the sugarcane plant remaining after the juice has been removed. Rice husks, a byproduct of cultivating one of the world's most widely grown food crops, is another example of agricultural waste that can be converted to energy.

Finally, urban waste-based fuels consist of various types of municipal wastes, including construction debris, furniture, yard waste (such as leaves and grass cuttings), old tires, and the methane that is produced in landfills. The potential contribution that can be made by municipal waste is small in comparison to the potential of the other two categories, but there are, as will be seen, good reasons to use it.

Another useful and common way of classifying biofuels uses only two categories, modern and conventional. Ethanol, biodiesel, synthetic gas, and black liquor are examples of modern fuels, because they are manufactured fuels or, in the case of black liquor, the byproduct of a manufacturing process. Conventional fuels include unprocessed wood, charcoal, and straw. The use of conventional fuels extends far back in history, but they remain important. Charcoal, for example, is still the main energy source for 200–300 million people.

Many fuels span both the modern and conventional categories, and manure is one of these. In regions where wood is scarce, some traditional societies continue to burn manure to generate heat. But manure is also a modern fuel because some farms and feedlots process manure to produce methane gas, which is burned for heat or power generation.

A third common and useful scheme classifies types of biomass, the raw material from which fuels are created, by the extent to which the materials have been changed from their original state. This scheme is somewhat similar to the conventional/modern classification scheme, but uses three categories instead of two. There are *primary resources,* which include fuel wood, crop residues such as the stalks and leaves left behind after harvest, grains such as corn, and specially grown grasses. *Secondary resources* include manufactured fuels such as ethanol and biodiesel and the byproducts of manufacturing processes such as black liquor and sawdust. *Tertiary resources* include the sorts of fuels that would be generated in more urban environments and that would pose disposal problems were they not burned and converted into energy. Tertiary resources include construction and demolition debris, municipal wastes, and landfill gases.

Multiple classification schemes reflect that fact that both the production and use of biofuels have numerous ramifications for society—ramifications that extend far beyond the energy sector. These issues include (but are not limited to) the environment, the cost of food, and ethics—whether, for example, it is ethical to use agricultural resources to produce motor-vehicle fuel when its production causes food prices to increase. In short, is it ethical to burn one's food? The issues associated with the production and use of biofuels tend to be more complex than those associated with other energy sources. Before addressing any of these issues, however, it is worth looking at an early history of fire and fuel.

A BRIEF HISTORY OF FIRE AND WOOD

Throughout most of history, in most areas of the world, wood has been the most important fuel. But while energy production has always been one important use for trees, it is also true that it has often not been the *most* important use. Trees have value as lumber, for example, as well as fuel. This situation is in sharp contrast to coal, which is produced solely to be burned. Coal can, therefore, be understood as a fuel, pure and simple. By contrast, living matter, because it has multiple uses, is a more complex resource. In addition to its value as a fuel source, for example, a stand of trees also has value—perhaps more value—as wildlife habitat, as a source of construction materials, as railroad ties, as telephone poles, or as a source of pulp for the papermaking industry, and many of these uses are mutually exclusive. A tree cannot, for example, be converted into both pulp and lumber, a nesting site for birds and a set of railroad ties.

But no matter the use to which a tree is put, part of the tree can usually be used for energy. A tree can, for example, serve as wildlife habitat even as dead branches are regularly removed for use as fuel, or a tree can be cut for lumber and the upper branches, leaves, and bark burned for their energy content. The preceding examples were chosen because they are modern and more familiar to the reader, but the same ideas applied to the use of trees in 15th-century Venice, ancient Rome, and in prehistoric cultures. The value of a tree is determined by the values of its owners, who can be expected to always choose the use that is of most value to them. Sometimes this involves burning trees for their energy content, but often it does not.

As a general rule, the value placed on wood is inversely related to the supply—that is, the greater the supply, the more is wasted. Consider the early agricultural settlements of central and northern Europe. Six thousand years ago, these settlements marked the path of a northward migration that had begun in the area around the

Tigris and Euphrates Rivers roughly 4,000 years before that, at the end of the last ice age. The culture is called the Neolithic, and the people of the Neolithic had farms and domesticated animals.

In Europe, early Neolithic settlers encountered dense forests. The forests made farming difficult and prevented the growth of the grasses their livestock depended upon for food. The forests were barriers to settlement. Although modern experiments have shown that it is possible to chop down a few trees with flint axe heads—and even as early as 6,000 years ago there was in Europe a lively trade in flint axe heads that spanned distances of hundreds of miles—the work involved in chopping many trees was prohibitive. These pioneers would almost certainly have saved their axes for cutting the wood they used to build the large wooden communal buildings in which they lived together with their animals. Instead, sections of forest were cleared by burning the trees where they stood. The ashes of the trees fertilized the soil. Fire applied to the undergrowth facilitated travel throughout the forest, and opening up the forest canopy facilitated the growth of feed for their livestock and forage for wild animals, which were another important source of food. There is evidence of repeated burnings of some forests during the Neolithic Age.

Firewood would also have been an important commodity in this cold climate, but chopping living trees would have been a poor way of procuring it. Even the most casual user of firewood knows that *green wood,* wood that has not been dried, is a poor source of fuel. It is difficult to burn, and because much of the thermal energy is wasted when the water in the wood is converted into steam, green wood generates a relatively small amount of heat. Instead, Neolithic settlers would have scoured the surrounding forest for deadwood. They may have found it on the forest floor or on still-standing trees. Dry deadwood burns readily and does not need to be "seasoned," or left to dry. Even though these early people lived in the heart of an enormous forest, their source of fuel was more restricted than their location would indicate. They would have scoured the forest, walking among enormous living trees, in search of dry firewood.

Kenyan women in the Rift Valley toting firewood. The production of modest amounts of heat can require large volumes of wood. *(Hans-George Michna; courtesy, AHEAD)*

What makes Neolithic practices "primitive" is not so much the tools that they used (Stone Age) but the market they served (themselves). Neolithic farmers were essentially subsistence farmers. What they produced, whether it was food, lumber, or fuel, was consumed on site by those who did the work. Although these agricultural practices are thousands of years old, they are still practiced by many people today. The widespread use of steel implements has, however, greatly increased the effects of small groups of people on large forests.

A more modern system for exploiting forest resources existed in ancient Rome, where 2,000 years ago, a sophisticated wood-fuel market had developed to serve the city's 1 million inhabitants. Romans used less wood for construction than many of their contemporaries, preferring instead to use bricks and ceramic tiles for many of their buildings. But before brick and ceramic can be used, they must be manufactured. In particular, they must be fired, a process that required large amounts of fuel wood. The mortar used to cement the bricks and tiles required lime, a building material manufactured from limestone by application of intense heat. It is estimated that manufacturing a ton of lime by the techniques then in use involved burning five to 10 tons of wood, and by 300 C.E., 3,000 wagon loads of lime were needed in the city each year for construction and maintenance. Space heating and Roman baths, which were common in this enormous city, required large amounts of fuel wood as well. Romans lived an energy intensive lifestyle. The less affluent used wood directly, and the more affluent used charcoal, which smokes less and provides a more even heat. Charcoal, however, also requires large amounts of wood to produce. Measured on a mass basis, six to 12 units of wood are consumed to yield one unit of charcoal, the difference in the charcoal yield depending partly on the skill of the producer.

Perhaps 90 percent of the wood that the Romans consumed was for fuel. Although estimates of the amount of wood required to build and heat Rome vary widely, they are all very large, and traders in fuel wood traveled far into the countryside in order to maintain the flow of wood into the city. (One conservative estimate, for example, is that providing fuel to Rome's 1 million inhabitants would have involved the consumption of roughly 12 square miles [31 km^2] of forest each year.) Some of this forest could have been replanted with seedlings, of course, to provide fuel for future needs, and some of it would certainly have been taken out of wood production permanently in order to create more farmland to furnish the city's residents with food. This illustrates (again) that there is more than one vitally important use for the land on which a forest is growing.

Commercial wood-fuel production and subsistence wood-fuel consumption often occur simultaneously. In 1750, the population of Boston, Massachusetts, was already in excess of 50,000, and wood was the fuel city residents burned to keep warm. The city depended on a far-flung network of suppliers. Local forests were already stressed as sources of fuel wood, and sailing ships brought fuel into Boston harbor from as far as 100 miles (160 km) up the New England coast. Transporting fuel wood across such distances greatly increased the price of the wood, and wood charities were created to help Boston's poor survive the winters. Faced with expensive wood and occasional supply shortages, urban residents began switching to the Franklin stove, named after its inventor, Benjamin Franklin, because the stove enabled the user to burn less wood while remaining just as warm. Meanwhile, only a little farther inland but away from commercial supply routes, rural communities were well-supplied with wood, often burning large amounts of it in open hearths, where, depending on the design of the hearth, as much as 90 percent of the heat escaped up the chimney—which again illustrates the point that given enough fuel the efficient use of wood is seldom an issue.

Stove design in the United States continued to improve. By 1860, 50,000 cast-iron stoves were being manufactured each year. But even with better stoves, wood was no longer competitive in urban areas. By the middle of the 19th century, large numbers of city dwellers were turning away from wood and toward coal for heat. In the east, anthracite, a type of coal produced in Pennsylvania with a high-energy output per unit volume, was the favored fuel. Not only was anthracite competitive economically, it was more convenient. As much heat is released burning a unit volume of anthracite as is released in burning a volume of wood that is five times as large, and in urban settings, where space is at a premium, the size of the fuel pile matters.

United States wood consumption peaked in 1907 at 13.4 billion *board feet* (31.2 million m³), a figure that includes all uses of wood, not just wood as a fuel. This was two-thirds of the world's output for that year. At the time, it was estimated that wood was being taken from the nation's forests at a rate that was three or four times greater than it was being replenished. Roughly 5 billion board feet (12 million m³) of that production was used for fuel. The volume of wood produced from the nation's forests has never again reached the levels attained during the early 20th century.

Large-scale fuel-switching—first wood to coal and later wood to other fossil fuels—began in the 19th century. Fuel-wood consumption probably peaked late in the 1880s, although estimates of the amount of wood burned as fuel necessarily involve a certain amount of guesswork, and experts disagree about the details. Nevertheless, wood remained an important part of the energy sector for the first few decades of the 20th century. When measured by the amount of thermal energy produced per year per fuel, wood fuel was still more important than natural gas as late as 1925. In the United States, wood use diminished slowly throughout the first half of the 20th century. But fossil fuel use increased quickly, and wood's relative contribution as a fuel source had become quite small by 1940, comprising only about 5 percent of the total thermal output of the United States.

Early in the 20th century, under President Theodore Roosevelt, the federal government began to put large tracts of forest into reserve. The Forest Bureau, the forerunner of today's Forest Service, was tasked with preventing an impending "wood shortage." There was a great deal of speculation about future timber "famines," but they never materialized. A new and underappreciated societal trend had already begun. Farmland was being abandoned, and forests were expanding again. The imbalance between the amount of wood grown and the amount consumed diminished, and by 1962, the Forest Service estimated that more wood was being grown than

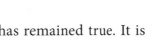

was harvested each year. For decades this has remained true. It is true today. The amount of wood in the United States continues to increase.

The United States was able to harvest wood at such a furious rate during the late 19th and early 20th centuries because its forests were so vast, and its population densities were, in most areas, very low. Other nations with higher population densities and smaller forests had to come to terms with the difficulties caused by poor for-est-management programs much earlier in their history. In many cases, however, the destruction of the forests in what are now called developed nations—destruction that occurred throughout much of the 18th and 19th centuries—has been largely reversed. Conserva-tion programs, increased imports of certain wood products, better forest management, the conversion of agricultural land to forest, and the cessation of attempts to create new farmland from forests have all served to preserve the remains of old forests and to create new ones.

The situation is often different elsewhere. Many tropical forests, for example, most of which are located in developing nations, are still shrinking. Part of the reason is that in some areas it is still com-mon practice to create agricultural land by burning forestland. Also, many small and large commercial interests operating in these areas are still taking trees without regard to the rate at which supplies can be replenished. Commercial interests often harvest specialty trees, such as mahogany, for the export market in ways that unnecessarily damage the entire forest. Small firms, sometimes consisting of one or few individuals, destroy trees to manufacture charcoal, a market that, in the developing world, still consists of hundreds of millions of consumers. When charcoal production is unregulated, the result is sometimes widespread destruction of forests.

There are those who assert that the experiences of developing countries are similar to those of developed countries and that de-veloping nations are just starting later. The developing world is a large place, and it is certainly possible to find situations where that statement seems to hold true, but in many areas, the situation is more complex. This is most evident in Brazil.

Two million people lived in present-day Brazil prior to the arrival of Europeans at the beginning of the 16th century. At that time the coast of Brazil consisted of an enormous forest. During the second half of the 16th century, three trends began to impact that coastal forest. First, sugar plantations, dependent on cheap land and enslaved workers, were established within the forests. This necessitated clearing large amounts of land. Even more land was cleared for firewood, which provided the energy needed to manufacture the sugar. (The liquids from the sugarcane must be boiled to produce the crystalline sugar. Roughly 3,600 cubic feet [100 m³] of wood was needed to produce 1.1 *short tons* [1 metric ton] of sugar.) Second, cattle, sheep, pigs, and goats were imported to create a new agricultural industry. Large sections of the coastal forest were burned in order to create grasslands on which the animals could feed. The development of coffee plantations led to the further destruction of the original coastal forests.

These efforts continued to expand until modern times. Agriculture and fuel-wood production pushed the boundaries of the forest back until the forest disappeared. It is estimated that in 1979, Brazil was meeting 79 percent of its energy needs with wood, a fact that reflects both the level of poverty in Brazil at that time and the absence of alternatives to wood fuel. A new trend began in the 1970s in response to the two oil shocks of that decade: Brazil initiated a program to produce ethanol from sugarcane as a strategy to reduce oil imports, a policy that required even more agricultural land to produce the additional sugarcane from which the ethanol was manufactured. Today, after centuries of exploitation, essentially all of Brazil's coastal forest—once one of the world's great forests and a resource that stretched along much of Brazil's long Atlantic seacoast—has been converted into agricultural land and grassland.

Brazil is now the world leader in exports of sugar and ethanol. In Brazil, ethanol is still produced from sugarcane, a crop that is, in many ways, ideal for ethanol production. And much of this sugarcane is still grown in what was once the great coastal forest. In addition to the ethanol produced for export, Brazil is also home to one of the most highly developed ethanol fuel markets in the world. Ethanol accounts for more than 40 percent of all fuel consumed by cars and light trucks. Where wood was once used to produce sugar, bagasse, the term used for the remains of the sugarcane plant after the juice has been removed, is used to produce much of the heat needed in the production of sugar and ethanol. Brazil has, as a matter of policy, traded forests for food, fuel, and coffee. Evidently, the Brazilian experience is long and complicated, but what it is not is a simple replay of the experiences of developed nations delayed by a century or two.

EARLY CONSERVATION EFFORTS

Until nations began to switch to fossil fuels, forests were a principal source of energy and raw materials. Forests were also often perceived as occupying land that might better be used for farming or grazing. As a consequence, forests were sometimes devalued and destroyed—but only sometimes. Centuries ago, some nations established successful programs to conserve or restore forests. The motivation for these programs was usually the perception of impending or actual wood scarcity and environmental destruction. Particularly noteworthy were early programs in France and Japan that successfully balanced human needs with good environmental stewardship.

The mountainous environment of southeastern France in the rural provinces of Dauphiné and Provence was in crisis by the middle of the 18th century. As the population had increased, more forested land had been cleared for farming, lumber, and fuel. Despite the

(continued on page 20)

Ancient Forests, Modern Humans

Today, it has become common to compare current forestry practices, good and bad, with an earlier time when people supposedly lived in harmony with nature and had only a minimal impact on the forest. This concept of "natural" is often used as a baseline against which to measure current efforts to exploit the resources of the forests: Managed forests are often viewed with suspicion and "wild" forests with approval. But this concept of a time when humans trod lightly over the Earth, a concept that influences many of today's discussions and decisions about the uses and values of forests and, in particular, the uses and values of biofuels, is a myth.

In the northern temperate zone, at least, many ancient forests are no more than about 10,000 years old. They began to develop at the end of the last ice age. As glaciers retreated northward, they left behind an environmental niche that was colonized by the ancestors of today's trees and today's humans. Although 10,000 years seems ancient, modern humans are considerably older. What this means is that in many parts of the temperate zone, for much of the past 10,000 years or so, humans and forests have evolved together, and there is a lot of evidence that humans created many of the characteristics of the forests that people today regard as "natural."

By far the greatest effects that technologically simple societies had on the forests in which they lived were the result of continual and intentional burning. There were, from the point of view of these early forest users, several advantages to continually setting the forest alight. First, by burning the underbrush, travel through the forest was made easier. Second, burning the forest deposited fertilizer in the form of ash, and seeds from fire-resistant and fire-dependent species on the ground. As a result, the ground was soon covered with tender shoots, which provided increased forage and so increased game. Third, the practice of firing the forest was valued as a form of pest control. Fourth, burning made it possible to more easily locate game by destroying places for animals to hide. Fifth, fire could be used to clear spaces for agriculture, and finally, fire was sometimes also used as a method of hunting. Fire-resistant species thrive in this environment; fire-susceptible species do not. The best accounts of the practice of firing the forest were written by European settlers and explorers during the 16th through 18th centuries. They were amazed by the extent of the burning that they witnessed in many places in the Western Hemisphere, and they were pleased with the results. In Virginia, many early European observers wrote admiringly about great open meadows, sometimes stretching for hundreds of acres, in which few or no trees grew. These grasslands had been created and maintained by Native Americans, among whom the practice of firing was widespread. Nor were these practices confined to the thirteen colonies or even North America. Captain James Cook in the journal he kept of his first (1768–71) voyage around the world wrote, ". . . and we have daily seen smokes on every part of the Coast we have lately been upon." These fires were set by Aboriginal Australians.

The resulting forests were not "natural" at all—not in the sense that they were the result of an absence of human intervention—nor was there anything harmonious about the method of their creation. In particular, the idea that all North American and Australian forests were unmanaged prior to the arrival of European settlers is certainly false. It is important to keep this in mind, because the production of significant amounts of biofuels will require the commitment of enormous amounts of agricultural resources and probably the intensive management of many forests. While there may be good reasons to decline to pursue such a program, the belief that it would disrupt the "natural" order is not one of them. For thousands of years, in many forests around the world, intensive forest management *was* the natural order, and the forests that resulted from these informal but intensive management programs are now characterized by many as a "natural" ideal.

(continued from page 17)

additional demands on the local forests, no serious attempt was made to replant the trees that were removed. Moreover, farm animals, left to browse on whatever vegetation they could find, ate the seedlings that did take root. Fast-diminishing forests were the result. As more trees were destroyed without replacement, wood fuel became impossible to find. Farmers and bakers alike began to burn dry cow dung to meet their heating and cooking needs. But the use of dung as an energy source meant a reduction in the amount of fertilizer applied to the land. Without trees to stabilize the increasingly exhausted soil, rains washed large amounts of soil into the valleys. So much soil washed down the mountains that it eventually impeded the flow of the rivers below. The soil that remained on the hillsides was less productive. Peasants left their ancestral homes in search of better lives elsewhere.

Faced with the environmental and social problems caused by deforestation, the government responded. Throughout the 19th century, the French government sought to reforest much of the denuded mountainous regions as well as the valley regions that had been severely impacted by soil deposition from above. Large tracts of land were brought under control of the government. Despite its good intentions, government efforts were not welcome, and many of the remaining inhabitants were initially resistant to the idea of government management. They were concerned that the new programs would interfere with traditional lifestyles, and they were reluctant to pay the costs of a very expensive program. By the middle of the century, however, many of these issues had been addressed, and reforestation efforts were under way on a large scale. Efforts were expanded as needed, and by the first decade of the 20th century the extent of forested land in France had increased by 20 percent, from 19 million acres (7.7 million ha) to 23 million acres (9.3 million ha). It remains one of the most successful reforestation programs in history.

In Japan, for a very long time, dependence on wood was complete. Throughout most Japanese history there were no alternatives to wood for fuel, because the country had no other fuel resources, and until the 19th century foreign trade was prohibited. Wood was also the only material used for construction. Demand for wood continued to grow as the population expanded. Spikes in demand were prompted by the need to rebuild large cities that were regularly razed by fires. The forests were disappearing. In Japan, attempts to control wood production for fuel and lumber and to manage the island nation's forests date back to the 17th century.

During the Tokugawa period in Japanese history (1603–1867), individual ownership of the forest was prohibited. Decisions about the forests were made by feudal rulers. In 1665, the ruler of Kiso, a district located on the island of Honshu, the largest island in the Japanese archipelago, issued an order to reduce timber production by half in order to give the forest time to regenerate. Seedlings were planted with the goal of reforestation, but the drop in timber production reduced the income of the area's ruler. Consequently, within a few years the restrictions were relaxed. The program failed. A second attempt was made in 1724. This time the goal was a 60 percent reduction in production, and this time production volumes were successfully reduced for 30 years, time enough to allow the forests of Kiso to regenerate. More modern techniques, including the taking of tree inventories in order to better plan future production, were put in place by 1779, a remarkable example of forward thinking. This system of forest management continued to evolve with the result that Japan remained forested even as its population continued to grow. It remains a well-forested island nation today.

The Japanese and French examples demonstrate that with respect to forests it is possible to choose both conservation and production. In fact, given enough time, later generations will often describe the forests that result from such policies as "natural." But not every crisis precipitates a creative solution. China's burgeoning population had denuded vast tracks of land by the 15th century, causing a fuel crisis that in some regions lasted centuries. For generations peasant farmers in some areas wandered the countryside for hours each day, collecting small amounts of flammable material wherever they could find it. On the other side of the world, the Onondaga nation of the Iroquois Confederacy in what is now the state of New York struggled continually with both food and fuel supplies. Their simple agricultural techniques caused farming productivity to drop over time as the soil's nutrients became exhausted, and the continual need for fuel wood, which was inefficiently burned, caused them to exhaust local fuel supplies as well. As a consequence, the capital of the Onondaga nation was moved at fairly regular intervals in order to access less depleted soil and live nearer more abundant supplies of firewood.

The fact that in large regions of 18th-, 19th-, and 20th-century China many peasants lived in a devastated landscape and spent hours each day searching for scraps of fuel is often blamed on poor forest management and high population densities. These were certainly contributing factors, but they do not explain the difficulties of the Onondaga in maintaining their fuel supply. They were, after all, a relatively small group of individuals living amidst a rich forest. To understand why fuel wood is difficult to harvest sustainably—that is, why wood, in particular, and biofuels, in general, are difficult to burn at a rate that does not exhaust the supply—it is necessary to look more closely at the physical and chemical characteristics of biofuels. In particular, in order to understand the value of biomass as a source of fuels, it is necessary to know how energy-rich a particular resource is. How much fuel is needed to produce a given quantity of thermal energy? How energy-rich are biofuels relative to other fuels? The answers to these questions reveal a lot about the ways that biofuels have been used and about limitations on their use today.

The Nature of Fuel

A fuel is only expensive or inexpensive, energy-rich or energy-poor, in relation to other fuels. This chapter considers the characteristics of some common biofuels. The first section compares these fuels to some other common fuels in terms of the amount of thermal energy released during combustion. The second section describes those factors that affect the cost and availability of biofuels and also describes ideas about the impacts associated with biofuel use. As with every fuel, biofuels must be understood in context, but compared to most other fuels, the context in which biofuels are used is more complex.

THE ENERGY CONTENT OF BIOFUELS

Biofuels are valued because they are sources of thermal energy. Sometimes this energy is used to heat a home or business, and sometimes it is used as fuel for a heat engine—a machine, such as

Switchgrass field—switchgrass is often touted as a fuel of the future.
(Mississippi State University)

a car or electrical generating station—that converts thermal energy into work. In any case, all other things being equal, the more thermal energy a unit of fuel produces when burned, the more valuable that fuel is. This raises the question of how much energy is released by burning a unit of a particular fuel.

Before the energy content of a biofuel is measured, it is important to be explicit about what is being burned. Roughly speaking, the amount of thermal energy released when burning a fuel sample depends on three important characteristics, the first of which is the "fuel component" of the fuel. The other two major components are water content and *ash* content. (Ash is the noncombustible solid component.) The fuel component of many nonwoody samples of biomass—materials as different as corncobs and rice husks—release roughly the same amounts of thermal energy per unit mass provided the water and ash have first been removed. But the water content of a fuel source can vary widely from sample to sample, even among samples produced from the same species of plant. Water content has a very large effect on the available thermal energy when a sample of biomass is burned. For example, wood that contains about 15 percent water, when measured on a unit-mass basis, will release about twice as much heat as green wood, which consists of about 50 percent water. Consequently, when discussing biofuels, engineers are always careful to identify the water content of the sample they have in mind. "Dry" fuel, also called *air dry*, contains about 15 percent moisture, because left in the air to dry, moisture content will not generally drop below 15 percent. Fuel that has had all of the water removed is often called "bone dry" or *oven dry*. Removing all of the water from a supply of biomass is energy intensive, and it is not usually done in practice. Most often, air dry must suffice.

The ash content of biomass can also vary widely. To return to the previously mentioned example of corncobs and rice husks, corncobs contain about 1 percent ash—that is, only 1 percent of a corncob, when measured by weight, consists of noncombustible solids. By contrast, about 15 percent of a rice husk is ash. Therefore, when equal weights of corncobs and rice husks are burned, considerably more heat can be generated from the corncobs than the rice husks, because less of the rice husk is combustible—that is to say, a significant fraction of dry rice husk is not fuel at all. In fact, in a sample of air dry rice husks, only 70 percent of the sample is combustible, the remaining 30 percent consists of water and ash. Ash further complicates the use of any fuel because from an environmental point of view, ash can represent a disposal problem. On a per-mass basis, therefore, air dry rice husks are, when compared to corncobs, an energy-poor and dirty fuel.

But if the preceding remarks help to identify why some biofuels generate more thermal energy than others, it says nothing in absolute terms about the amount of thermal energy a given fuel releases when burned. The energy content of a particular fuel is usually expressed either as the amount of thermal energy released per unit mass or in terms of thermal energy released per unit volume. Many science books emphasize the importance of measuring things on a per-mass basis, because mass is more "fundamental" in the sense that the mass of a sample of material will not change if the material changes location or phase (from the Earth to the Moon or from a solid to a liquid, for example). But whatever the value of "per-mass" measurements from a scientific perspective, engineers and scientists in the field of transportation fuels often prefer to talk about energy content per unit volume.

Although mass may be more fundamental, volume is often more relevant. The amount of space in a car or plane that can be devoted to carrying fuel is very limited, so the volume (not the mass) of the fuel is the critical characteristic. A fuel that requires a large storage volume would probably not be practical to use in the transportation sector, because of limitations on the amount of space that can be devoted to storage. By way of example, the energy content of ethanol is usually expressed as 75,700 *Btu* per gallon (21.1 MJ/l), and the energy of gasoline is about 116,000 Btu per gallon (32.2 MJ/l). A car that ran on pure ethanol would, therefore, require a fuel tank that was 50 percent larger than that of an otherwise identical car that ran on pure gasoline in order to have the same cruising range, and a larger tank means less storage and passenger room.

The energy content of gaseous fuels is usually expressed in terms of the amount of thermal energy released when burning a unit volume of gas at "standard" temperature and pressure, which in this book will mean 60°F (16°3C) and one atmosphere pressure. The energy content of natural gas, for example, is about 930 Btu per cubic foot (34.6 MJ/m^3) at standard temperature and pressure. Different systems of measurement are used for different types of fuel in order to better reflect the physical properties of the fuel in question as well as the conditions under which it is stored and used.

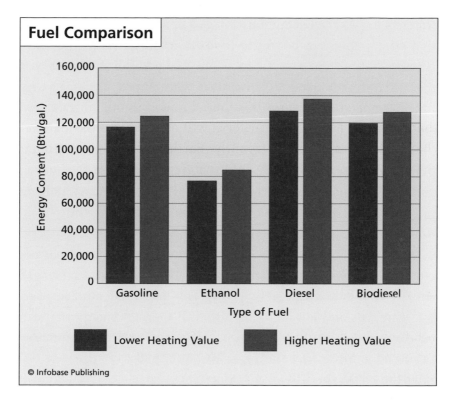

Fuel comparison. There is much less energy in ethanol than in the same volume of gasoline, but equal volumes of biodiesel and conventional diesel fuel yield almost equal amounts of thermal energy.

Finally, it is important to know that there are two common ways of expressing energy content. Sometimes the energy content of a fuel is expressed in terms of the *higher heating value* (HHV), sometimes in terms of the *lower heating value* (LHV). The difference between the two heating values depends on when the measurement is made.

Burning any biofuel produces mostly thermal energy, carbon dioxide (CO_2), and water. (This water is in addition to the water that was already present in the fuel prior to combustion.) The water that is created during the combustion reaction is initially in the form of water vapor. As the hot combustion gases expand and cool, some of that water vapor may condense to form liquid water. As the water changes phase from a vapor to a liquid, a good deal of energy is released into the surrounding medium without producing a change in the temperature of the water. One can measure the thermal energy produced by the combustion reaction before the water changes phase or after. If one measures the thermal energy before the vapor-to-liquid phase change, one gets the LHV. This is common practice for transportation fuels, because heat engines generally do not convert the energy released during the phase change into work. But sometimes the measurement is not made at the same temperature as the LHV measurement but after the phase change has occurred. This measurement might be made when the fuel is burned simply for the thermal energy it releases—in a residential furnace, for example—and not for the amount of thermal energy that can be converted to work. If the measurement of the thermal energy produced is made after the vapor-to-liquid phase change, one obtains the HHV. The difference between the two values is not large in comparison to the amount of heat released during combustion, but it is important to be aware of the difference since both methods of expressing energy content are in use. (By way of example, biodiesel fuel releases 118,296 Btu per gallon [35.133 MJ/l] LHV and 128,520 Btu per gallon [38.170 MJ/l] HHV.)

AVAILABILITY AND COST

How much of a contribution can biofuels make? This question is surprisingly difficult to answer, because often there is a big difference between the amount of biomass that exists—biomass is the raw material from which a biofuel is manufactured—and the amount of biomass that is available for use as a fuel. This reflects two facts about biomass that will appear repeatedly in the chapters that follow.

First, as previously mentioned, biomass usually has more than one use. Dairy farmers who produce electricity from cow manure,

In the United States, most enthanol is produced from corn. *(iStockPhoto.com)*

for example, have less fertilizer to spread on their fields. Because their fields need to be fertilized, they must balance their need for fertilizer with their need for heat and power. Similarly, corn farmers who sell their crop to produce ethanol have effectively taken their land out of production with respect to food and feed. Decisions about how much land to devote to ethanol production versus food and feed production is informed in part by the price of ethanol versus the price of food and feed. Variations in the profits generated by producing each product affect future production decisions, which in turn affect future profits. These decisions are difficult and economically risky. There are many other examples of how producers must balance the multiple uses to which a particular agricultural product or byproduct can be put, some more subtle than others. What is important to understand is that the production of any biofuel feedstock represents just one possible use of biomass from among several alternatives, and no matter which alternative one chooses, there are serious economic and environmental consequences.

The second fact about biofuels that strongly affects their availability and cost is that, while there is plenty of biomass, it is often dispersed over a wide area. To see how this affects its value, consider the problem of determining how much energy can be obtained from soybeans, which, in the United States, are an important feedstock for biodiesel, a biofuel that can be used in place of diesel fuel and home heating oil. Farmers in the United States have planted soybeans across enormous expanses of land. In some places the fields stretch right to the horizon—but how much energy does a field of soybeans represent?

Currently, the average yield for soybeans is about 2.4 short tons of beans per acre (5.3 metric tons per hectare), which translates into an oil yield of about 48 gallons per acre (450 liters per hectare). This is a very low yield in the following sense: Forty-eight gallons of liquid evenly dispersed over an acre of land would form a film 0.0018 inches (0.0044 cm) thick, roughly the thickness of a human hair. But this energy is not there for the taking. It must be harvested. On a large farm, harvesting involves, among other things, using large energy-hungry harvesters. The farmer must also transport the soybeans to a processing plant. The process by which biodiesel is manufactured requires additional energy. All of these activities drive up costs and diminish the net value of the biodiesel produced. But these activities are also unavoidable, because the soybeans are spread across the landscape. In fact, biodiesel produced from soybeans is better characterized as a coproduct of soybean production. The use of soybean oil to manufacture biodiesel contributes to the profits enjoyed by the farmer, but these profits are not large enough to justify the cost of growing the soybeans. The main value of soybeans is in their value as food and feed.

With respect to soybeans, it might appear that one way to increase the biomass yield would be to use the rest of the soybean plant. Although the stems and leaves of the plant have no value with respect to the production of biodiesel, they do, at least, constitute a large source of biomass and so a potential source of fuel. But the stems and leaves, called the residue, are unavailable for energy production. Good conservation practice requires that the soybean residue be left in the field in order to inhibit erosion and maintain soil fertility. Therefore, although it initially seems as if the residue represents an important source of biomass, no part of it is available for energy production.

The preceding examples illustrate a much larger question: How much biomass is available to use at a cost people can afford? This was discussed in a well-known and widely quoted 2005 paper published by the United States Department of Agriculture (USDA) entitled "Biomass as Feedstock for a Bioenergy and Bioproducts Industry: The Technical Feasibility of a Billion-Ton Annual Supply." The purpose of the paper was to examine whether sufficient biomass could be gleaned from within the United States to displace 30 percent of the nation's petroleum consumption by 2030. The conclusion of the paper was that it was possible to offset 30 percent of the nation's current oil consumption with biofuels of various types, but because the paper was written in a way that lends itself to being misquoted, it is important to be clear about what this means.

When the paper was first published, biofuels constituted only about 3 percent of the nation's energy consumption. They displaced a similarly miniscule percentage of the country's petroleum consumption, so there was no possibility of biofuels having a dramatic impact on the nation's energy supply at that time. And petroleum consumption does not remain constant. On average, it increases over time. This trend has held true for many decades, and the United States Energy Information Administration (EIA) expects that consumption will continue to rise at least through 2030. Consequently,

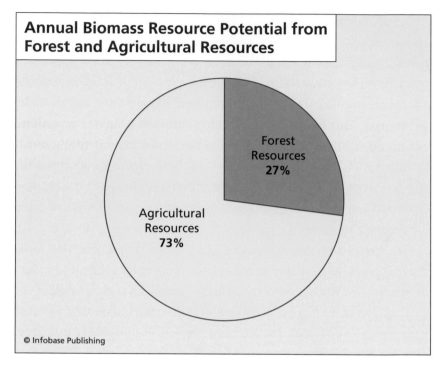

Annual Biomass Resource Potential from Forest and Agricultural Resources

Forest
Resources
27%

Agricultural
Resources
73%

© Infobase Publishing

A large biofuel sector will be heavily dependent on agricultural resources.
(Source: USDA)

even if a biofuels program were pursued very aggressively and the goals in the paper were met, it would not displace 30 percent of the nation's petroleum supply in 2030 because the United States will be using more petroleum then than it was in 2005—an estimated 20 percent more. Thirty percent of today's market translates into 25 percent of the projected market. Some of the biomass would be used to displace some of the petroleum used as feedstock by the petro-chemical industry—that is, it would displace some of the petroleum from which many materials are manufactured—and some of the biomass would be used to displace some of the nation's transportation fuels. Some biomass might be used for power generation, but it is doubtful this would displace much petroleum in 2030, since the amount of oil burned to produce electricity has been dropping since the oil crises of the 1970s. Currently, only a small percentage of the nation's electricity is produced by burning oil.

Displacing 30 percent of today's oil supply would require in the neighborhood of 1 billion dry short tons (900 million metric tons) of biomass. This figure did not include biomass located in forestland not accessible by road or biomass located in environmentally sensitive areas. The authors considered this biomass to be unavailable. Their estimate also did not include biomass located far from any potential market, because presumably, this biomass would be too expensive to gather. They assumed that the yield per acre of corn, wheat, and other grains would increase 50 percent by 2030. (Historically, yields on these crops increase regularly.) They assumed that soybean plants would be developed with much larger amounts of residue, so that some of the residue would be harvestable. They assumed animal manures not needed for fertilizer would be used for energy production, and they assumed a number of other changes in agricultural practices favorable to the increased production of biomass. Evidently, this entails many optimistic assumptions, but they concluded that if all of these conditions were met, the nation could responsibly produce 1.3 billion dry short tons (1.2 dry metric tons) of biomass, enough to meet 30 percent of the United States' petroleum requirements in 2005 or perhaps 25 percent of the nation's requirements in 2030.

Forestry resources, including primary fuel wood from forests, secondary sources generated at lumber and paper mills, and tertiary sources such as construction and demolition debris, would generate 368 million dry short tons (334 dry metric tons) each year and the rest, which would amount to 998 million dry short tons (905 dry metric tons), would come from agricultural sources, primary, secondary, and tertiary. To produce so much biomass would entail an enormous commitment of resources. Because studies indicate

(continued on page 36)

Sinks, Sources, and the Greenhouse Effect

"I worked out the calculation more in detail, and lay it now before the public and the critics."

> —*Svante Arrhenius (1859–1927), describing the greenhouse effect for the first time in his 1898 article, "On the Influence of Carbonic Acid in the Air upon the Temperature of the Ground."*

The term *greenhouse effect* began as an analogy between Earth's atmosphere and the glass walls and roof of a greenhouse. On a sunny day, sunlight will stream into a greenhouse. Light is a form of energy that passes readily through glass. When a ray of light passes through the glass and strikes a plant, a table, or other object inside the greenhouse, some of the light is reflected back out through the glass. Light, having passed easily through the glass on its way in, passes easily through the glass on its way out. But some of the light is absorbed by objects inside the greenhouse, living or not, and then radiated back into the interior of the greenhouse in the form of heat.

Both heat and light are forms of energy, but light easily passes through glass, and heat does not. This difference between the two forms of energy explains why greenhouses are so warm on sunny days. Energy enters quickly in the form of light but leaves slowly in the form of heat, causing thermal energy to accumulate within the greenhouse and resulting in an increase in greenhouse temperature. The type of glass used to build the greenhouse also affects the ways that light and heat pass into and out of the greenhouse. Some types of glass insulate better than others.

Air is largely transparent to sunlight. Light streams through the atmosphere and strikes plants, oceans, rocks, and soil. Some of this light is reflected, and some is absorbed, and the energy is radiated back into the atmosphere as heat. As with glass, the atmosphere is not transparent to heat, and thermal energy accumulates in the atmosphere, causing the temperature to increase. The rate at which heat accumulates is determined in part by the level of carbon dioxide (CO_2) in the air. The more CO_2 that accumulates in the atmosphere, the more heat is retained, a fact that

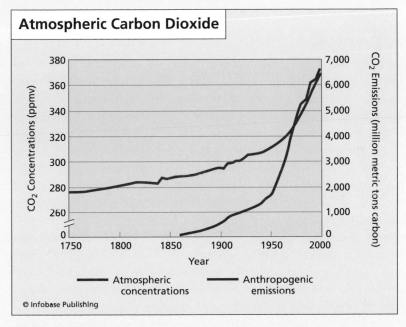

The concentration of atmospheric carbon dioxide has increased in step with increased carbon dioxide emissions due to human activity. *(Source: EIA)*

explains the analogy between a planet and a greenhouse. Changing the amount of CO_2 in the atmosphere is analogous to changing the type of glass used in the greenhouse.

Earth's atmospheric chemistry changes continually. Large amounts of CO_2 cycle into and out of the atmosphere. The oceans absorb billions of tons of CO_2 each year, and plants absorb CO_2 as well. Major emitters of CO_2 include forest fires, decaying vegetation, automobiles, and fossil fuel–fired power plants. Carbon dioxide emitters are called *carbon sources,* and those objects that absorb CO_2 are called *carbon sinks.* There

(continued)

(continues)

have been times during Earth's history when sinks have been more active than sources, causing a drop in CO_2 levels in the atmosphere. Other times, sources emitted CO_2 more rapidly than the sinks could absorb it, causing CO_2 levels to rise. Such is the case today. As CO_2 levels rise, average global temperature increases.

Biofuels are different from fossil fuels in that they represent sources *and* sinks. As biomass accumulates, it functions as a sink, drawing CO_2 out of the atmosphere and incorporating the carbon into the tissues of the plant. Growing biofuel feedstock, then, helps to reduce atmospheric CO_2 levels. When biofuels are burned, CO_2 is emitted back into the atmosphere. By contrast, burning fossil fuels emits CO_2, but fossil fuels do not form fast enough to be considered sinks.

It is sometimes claimed that biofuels have no net impact on atmospheric CO_2 levels, because they act as both sources and sinks. This is false. The production of biomass partially offsets the CO_2 emitted during their combustion, but substantial quantities of CO_2 are also emitted when fertilizer is manufactured, when the fertilizer is applied, when the seeds are sown, and when the plants are harvested, transported, and converted to biofuels. The effects of biofuel production and consumption are complex and not yet fully understood.

(continued from page 33)

that increases in biofuel production have already caused increases in food prices, biomass production on the scale envisioned by the authors of the study would probably also result in a number of additional changes to the way Americans live.

This paper, which is thoughtful, carefully reasoned, and extremely optimistic, can be viewed as placing an upper limit on the capacity of the U.S. biofuel sector to meet demand. To put its conclusions another way: The best that can be expected from biofuels is a decrease in the projected demand for petroleum in 2030 by (at most) 25 percent. This is a significant reduction, but even under the best of circumstances, it is apparent that biofuels are no substitute for oil. As will be seen, however, there are other good reasons for aggressively developing the biofuels market.

BIG AND SMALL ENVIRONMENTAL AND ECONOMIC EFFECTS

Biofuels are sometimes portrayed as environmentally benign. At current rates of production, there are only modest environmental benefits or drawbacks associated with biofuels use. At present, they are not used intensively enough to make a big difference in fuel consumption patterns or in the environment as a whole. But the positive and negative aspects of biofuels, as well as broader questions about their value, are becoming more apparent as the scale of production continues to increase.

The United States farming economy has changed enormously since the late 1950s and early 1960s. Using the most up-to-date equipment and the most productive hybrids, 1960s-era American farmers found that they could produce more food than ever before. Americans ate better and spent less on food than their counterparts in most other nations. As world economic conditions changed, American agricultural exports increased rapidly. Although exports had been increasing since the 1950s, they surged throughout the 1970s even as enormous domestic surpluses were maintained. It was often said that American farmers could feed the world, and there was some truth in the saying. But if the world had discovered that it could rely on American farmers for sustenance, American farmers discovered that even after feeding the world, they still had some left over. Continual yearly surpluses and increasing foreign

This enormous pile of corn illustrates the extraordinary scale on which modern agriculture operates. *(iStockPhoto.com)*

competition resulted in downward pressures on grain prices, and during the 1980s many farmers found it impossible to continue to earn a living in agriculture. Large areas of the Midwest fell on hard times.

As demand for ethanol and biodiesel has climbed, so have prices for the biomass feedstocks from which they are produced. Prices have climbed even as more land has been brought into production. There is a new and welcome prosperity in many agricultural communities that is due to the increased market for biofuels. In addition, farming is a knowledge-based industry, and farmers, as practitioners, are some of the most knowledgeable individuals in the industry. By preserving farms, the knowledge necessary to farm productively is also preserved. Finally, agricultural land is a national resource. Preserving agricultural land for production rather than for other types of development is best done by keeping the land in production. In these senses, a prosperous agricultural sector is good for the nation.

On the negative side, large-scale production of biofuels can have a number of complicated and unintended effects. Recall that in the USDA study described in the previous section, the authors estimated that in order to reach their biomass production target, almost three-fourths of the biomass would have to come from the agricultural sector. This is a tremendous commitment of agricultural resources. Although the authors are confident that this level of production can be achieved without food shortages, they are less clear about the effects of so much biomass production on food prices. The distinction is, in any case, a fine one. For those less well-off, a hike in prices has the same effect as a food shortage.

There is ample evidence that as more land has been devoted to producing corn for ethanol production, food prices have risen, but the relationship between increased ethanol production and increased food costs is controversial. Energy prices are also an important agricultural cost, and oil and natural gas prices have increased as ethanol production has climbed, a fact that has complicated the analysis. Nonetheless, some facts are worth recalling.

First, corn is a staple. Unprocessed, it is a common side dish. Processed, corn is a common ingredient in a wide variety of foods, and other corn products, especially high fructose corn syrup, are used in the preparation of everything from candy to bread. The price of these items must reflect in part the price of corn. Less obvious, but more significant, is the effect that corn prices have on meat and dairy prices. Because corn is a major ingredient in animal feed, increases in the price of corn have also translated into increases in the price of milk, eggs, and meat. (During the first 10 months of 2007, for example, milk prices rose 18 percent, although, again, this cannot be attributed solely to the cost of corn.) Finally, many farmers who want to increase corn production do so by planting more

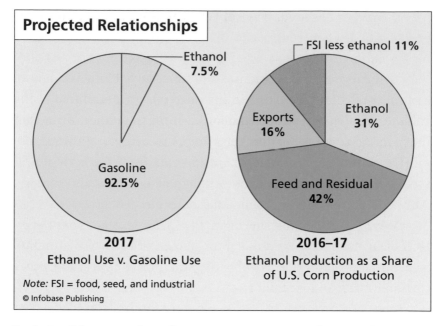

Projected Relationships

Ethanol
7.5%

Gasoline
92.5%

2017
Ethanol Use v. Gasoline Use

FSI less ethanol 11%

Exports
16%

Ethanol
31%

Feed and Residual
42%

2016–17
Ethanol Production as a Share
of U.S. Corn Production

Note: FSI = food, seed, and industrial
© Infobase Publishing

Replacing 7.5 percent of gasoline consumption in 2017 will require almost one-third of the corn grown in the United States. *(Source: USDA Agricultural Projections, February 2007)*

corn at the expense of other crops. The shift in agricultural land usage has meant that as corn production increased, the production of some other food and feed crops has decreased, and food prices have risen as a result. (In 2007, for example, soybean prices climbed 40 percent, an increase attributed in part to a reduction in the acres of soybeans planted in favor of corn.) Although opinions differ on the relative roles of increased corn production, changing land-use patterns, and increasing oil and natural gas costs on food prices, it is a certainty that grain prices will remain volatile for years to come as energy and food and feed markets evolve together. Increased yields, an expanding agricultural sector, and the price and availability of petroleum and natural gas will all affect the way that the food and biofuels sectors develop. There are, in any case, real costs associated

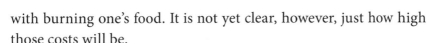

with burning one's food. It is not yet clear, however, just how high those costs will be.

In addition to large-scale effects associated with large markets, there are also an increasing number of small-scale biofuels projects. In fact, some biofuel projects, a few of which are described in this volume, are operated on such a small scale that one could argue that they are too small to matter. Yet some of these same projects perform very important functions, *but energy production is not one of them.* The reason is that some kinds of biomass are not just energy sources—they are environmental hazards.

Consider, for example, the problem of disposing of cow manure, a problem faced by many dairy farmers. At one time, dairy farmers spread manure over their fields until the accumulated supply was exhausted. This practice solved two problems: It fertilized the fields, and it disposed of the manure. In some northern regions, however, farmers are no longer permitted to spread manure across their fields during the winter months, because bacteria are unavailable to decompose it. Experience has shown that the manure remains on the field until the spring, when melting snow and rain wash the "winter manure" into lakes and streams, accelerating the growth of algae, sometimes creating serious pollution problems.

Manure spreading remains an important part of soil management, but because of the restrictions on the spreading of manure during the winter months, some farmers are left with accumulating supplies of waste that, over time, may pose an environmental hazard even if they are not spread. These wastes certainly constitute a nuisance. In response, some farmers have installed digesters, devices that convert much of the surplus manure to a biogas that can be burned to produce either heat or electricity. On small farms, these are small projects, and they are necessarily inefficient because they cannot benefit from economies of scale. The proof of this is that many farms do not install digester technology without taxpayer-subsidized grants. But if the contribution of the digester-based heat and power system to the regional energy supply is negligible, its contribution to the environment is not, because a small system solves 100 percent of the farmers' manure problems. There are many similar applications of biofuel technology that do more for the environment than their contribution to heat and power production would indicate. Small energy projects do not necessarily have small environmental effects.

The Chemistry
of Combustion

The combustion of biomass can be understood as the inverse of photosynthesis. During photosynthesis, plants combine energy from the Sun with carbon dioxide and water to produce biomass. Combustion reverses the process, because it combines oxygen with biomass to produce carbon dioxide, water, and thermal energy. It is a simply stated process, but as a practical matter many biomass combustion systems have proven difficult to operate efficiently.

This chapter covers these three topics.

> First, it provides some theoretical background on the combustion reaction. A theoretical combustion reaction provides insight into the relation between what is burned and what is produced as a consequence of the combustion process.

> Second, practical considerations related to biofuel combustion are described.

> Third, the chapter closes with an explanation of how thermal energy is converted into work.

Biofuels are extremely diverse in terms of their characteristics and their uses, but commonalities are also important. This chapter seeks to identify some common characteristics.

THEORETICAL COMBUSTION

Three elements are present in large quantities in biomass: carbon, hydrogen, and oxygen. When discussing the theoretical combustion of biomass, it is usually assumed that these are the only elements present. That assumption is false. Most biofuels are chemically more complex than theoretical considerations indicate. But because carbon, hydrogen, and oxygen are the main elements participating in the combustion reaction, assuming that they are the only constituents greatly simplifies the combustion model while still revealing some important aspects of combustion.

There are two other assumptions that are often made in modeling the chemistry of a combustion reaction. First, it is assumed that all of the carbon present in the fuel combines with oxygen to produce carbon dioxide (CO_2), and second, it is assumed that all of the hydrogen present in the fuel combines with oxygen to produce water (H_2O). A fuel that burns according to these assumptions is completely combusted. It is clean-burning in the sense that a combustion reaction that produces only CO_2 and H_2O is, environmentally speaking, the best possible outcome. Although the release of large amounts of CO_2 into the atmosphere can have serious environmental consequences, any outcome other than the production of CO_2 and H_2O will have environmental effects that are even worse.

Combustion reactions are sometimes described with chemical equations. These equations are important, because they unequivocally identify what is burned during combustion and what is

The 11.8 teraflop Hewlett-Packard supercomputer at Pacific Northwest National Laboratory has more than 2,000 processors and 6.8 terabytes RAM. It is used for combustion calculations. *(Department of Energy Archives)*

produced by the combustion reaction. A chemical equation reveals the relative amounts of fuel and oxygen required for complete combustion. Broadly speaking, a combustion equation will have five terms written in the following way:

$$\text{fuel} + \text{oxygen} \rightarrow \text{carbon dioxide} + \text{water} + \text{heat}$$

The arrow in the chemical equation indicates the direction of the reaction. One begins with fuel and oxygen and ends with carbon dioxide, water, and heat.

The oxygen in a combustion reaction is obtained from the air, and in the air, oxygen atoms are present in the form of oxygen molecules, each of which consists of two oxygen atoms bound together. Symbolically, molecular oxygen is written as O_2, the subscript "2" indicating the number of atoms of the element present in the molecule.

Sometimes a simple chemical formula for the fuel is available. When such a formula is available, it can be used in a chemical equation to specify the relative number of reactant molecules needed for complete combustion as well as the relative number of product molecules produced as a result of the reaction. These numbers, called coefficients, are chosen so that the same numbers of carbon, hydrogen, and oxygen atoms that appear on the left side of the equation also appear on the right. That the number of atoms remains unchanged by the combustion reaction reflects the fact that chemical reactions do not alter atoms, only molecules.

To see how this can work in practice, consider the combustion of ethanol, an automotive fuel made from corn. The equation for an ethanol molecule (C_2H_6O) shows that it is composed of two carbon atoms, six hydrogen atoms, and a single oxygen atom. (The absence of a subscript means that there is only a single atom of that kind.) The idealized combustion of ethanol is represented by the following chemical equation:

$$C_2H_6O + 3O_2 \rightarrow 2CO_2 + 3H_2O + \text{heat} \qquad \text{(Equation 3.1)}$$

Equation 3.1 shows that to burn one ethanol molecule completely requires the presence of three oxygen molecules. It also reveals that the combustion of the ethanol molecule produces two molecules of carbon dioxide and three of water. Evidently, the combustion of ethanol results in the production of a good deal of carbon dioxide.

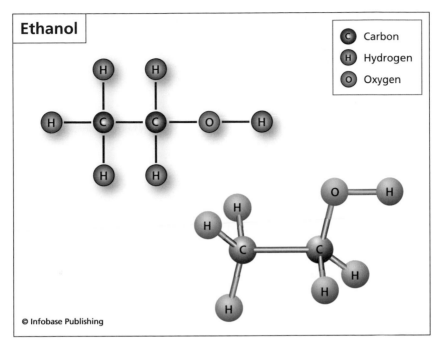

Ethanol

C Carbon
H Hydrogen
O Oxygen

© Infobase Publishing

Two representations of an ethanol molecule

It is also possible to write the amount of heat released by the combustion of a single ethanol molecule, but that number is very small, and its description would take the present discussion too far afield. What is important to understand here is the reason that heat is released at all. Energy is required to bind together the atoms that form the ethanol and oxygen molecules on the left side of the equation. Similarly, there is a certain amount of energy involved in binding the atoms that form the carbon dioxide and water molecules on the right side of the equation. But the amount of energy required to produce the molecules on the left side is larger than the amount required to produce the molecules on the right. Since energy is neither created nor destroyed, there is energy "left over" after the production of the carbon dioxide and water molecules. This energy appears as the heat produced by the combustion reaction. Put another way: The heat produced during a chemical reaction is the difference between the chemical energy of the reactants and that of the products.

Among biofuels, ethanol is unusual in that it is a chemically homogeneous fluid. Pure ethanol consists solely of molecules of C_2H_6O. (Ethanol is sometimes represented by a closely related formula, C_2H_5OH, which emphasizes the presence of the *OH* group of atoms. This second formula still shows that two carbon, six hydrogen, and one oxygen atom comprise the ethanol molecule.) Because ethanol consists of a single type of molecule, it is possible to write a single chemical equation for ethanol combustion. Most other biofuels are more difficult to characterize, because they are not chemically homogeneous.

Biodiesel, a fuel that can substitute for diesel fuel or home heating oil, is an example of a more chemically complex fuel. It is made from a variety of feedstocks, including grease from deep fryers, sunflower seeds, and soybeans. Not surprisingly, the chemical composition of biodiesel depends somewhat on the feedstock from which it is manufactured. As a consequence, there is no single "biodiesel molecule" in the sense that there is a single ethanol molecule. Nor does a sample of biodiesel fuel usually consist of liquid produced from a single feedstock. The chemical composition of biodiesel fuel varies, therefore, from place to place. The precise chemical composition can also vary from day to day, as supplies are replenished with different biodiesel blends. These variations are invisible to the consumer, because different types of biodiesel burn in essentially the same way. Different samples have the same heating value, for example; they will ignite under the same conditions, and their combustion produces approximately the same sorts of products. This situation is summarized by saying that the "bulk properties" of biodiesel are similar no matter which feedstocks were used in the production of a particular sample.

The combustion of wood is even more resistant to a simple chemical equation–type of analysis. As with biodiesel, there is no single "wood molecule," no formula that one can write in place of the formula for the ethanol molecule appearing in the preceding combustion equation (see Equation 3.1). But in contrast to biodiesel, the bulk properties of wood are not uniform. They vary greatly from sample to sample. The density of wood, its heating value, its moisture content, and its ash content depend on the species of tree to which the sample belonged, the location of the tree, and the soil and moisture conditions prevalent at the site. Nor can the bulk properties of wood be made uniform by processing. Wood is, in a sense, not one fuel but many.

Despite these complications, the theoretical combustion of biodiesel and wood (and other biofuels as well) has much in common with the theoretical model of combustion of ethanol. The complete combustion of these fuels produces, for the most part, carbon dioxide, water, and heat in quantities that are characteristic of the fuel being burned. The difference between the combustion of ethanol and most other biofuels is that the properties of the other fuels can only be described using average values—for example, the average amount of heat released per unit of fuel combusted and the average amount of carbon dioxide produced per unit of fuel.

Average values are determined by experiment. They are obtained in a laboratory, where representative samples of a fuel are carefully burned with sufficient quantities of oxygen to maximize the production of carbon dioxide and water. Precise measurements are made of the products of combustion—that is, measurements are made of the ratios of the masses of the products, for example, and the amount of thermal energy produced per unit of fuel consumed. The data are analyzed statistically. This type of analysis enables scientists to make various important deductions about the energy content of a fuel and the amount of oxygen required to completely combust a sample of the fuel even without a chemical equation, such as Equation 3.1. Conclusions are often expressed in the language of probability. An engineer or scientist may say, for example, that the heating value of a randomly selected sample of a particular type of biofuel will be located within some experimentally determined range of heating values at least 95 percent of the time, or that the amount of carbon dioxide produced from the combustion of one unit of fuel will fall within a particular range of carbon dioxide values 99 percent of the time.

The key to this way of characterizing a fuel is multiple repetitions of the same basic experiment followed by rigorous analysis. By repeating a particular experiment numerous times and statistically analyzing the resulting data, combustion engineers and scientists are able to quantify their uncertainty about the bulk properties of a heterogeneous fuel. Although these probabilistic statements are very different from the very deterministic type of statement expressed in Equation 3.1, from a practical point of view, they are often just as useful.

COMBUSTION: SOME PRACTICAL CONCERNS

Broadly speaking, each biofuel generates less thermal energy than the fossil fuel to which it is most similar. Landfill gas, for example, has a heating value that is usually between one-third and two-thirds that of natural gas. Ethanol generates only about two-thirds as much thermal energy as the same volume of gasoline. Biodiesel generates slightly less thermal energy—between 5 and 10 percent—than the same volume of diesel fuel, and wood usually generates less thermal energy per unit mass than coal. (The physical properties of both wood and coal vary a good deal, and there is some overlap—that is, under some conditions some samples of "good" oven dry wood may generate more thermal energy than some "inferior" samples of coal with high moisture content—but most coal samples yield significantly more heat than most wood samples.)

Although most biofuels are energy-poor relative to their fossil fuel counterparts, other considerations add to their value. In par-

Brookhaven National Laboratory has developed a more efficient oil burner. Here it is burning a biodiesel-petrodiesel blend. *(Department of Energy Archives)*

ticular, the production or consumption of biofuels is often accompanied by the production of coproducts, materials that have value other than as energy sources. A few examples have already been mentioned. Here are two more: The manufacture of ethanol fuel produces animal feed as a coproduct, and depending on the method by which ethanol is manufactured, other coproducts are produced as well. (See chapter 4 for information about ethanol production methods.) The gasification of *poultry litter,* the process that converts wastes from large-scale poultry farms into a gaseous fuel, produces fertilizer as a coproduct. (See chapter 6, "A Case Study: Power Generation versus the Environment.") Alternatively, there are often environmentally compelling reasons for the combustion of a biofuel that have little to do with energy production. Removing dead trees and branches from a forest for use as fuel wood, for example, increases energy supplies and decreases the risk of out-of-control forest fires. Harvesting deadwood is, then, a valuable forest management tool as well as a means of procuring fuel. (See chapter 8, "Fire Suppression.") Many other examples exist, some of which will be discussed later in this volume. From a practical point of view, therefore, most biofuels cannot be judged on their heat content alone.

Another characteristic that is shared by most biofuels is that their production methods are both labor- and energy-intensive. With respect to energy crops, such as corn, rapeseed (also known as canola), and switchgrass, this is obvious. Crops must be planted, tended, harvested, transported, and processed prior to use. The same general pattern is also true for other biofuels. These general statements can be made to sound more precise by examining the ratio of energy output to energy input—that is, the ratio between the amount of thermal energy produced during the combustion reaction to the amount of energy expended manufacturing the fuel. Although there is some controversy with respect to the way that energy input should be measured, all analyses indicate that the energy-output-to-energy-input ratio is lower for most biofuels than for the fossil fuels they most closely resemble.

With respect to corn-based ethanol, part of the debate even concerns whether more energy is expended producing ethanol

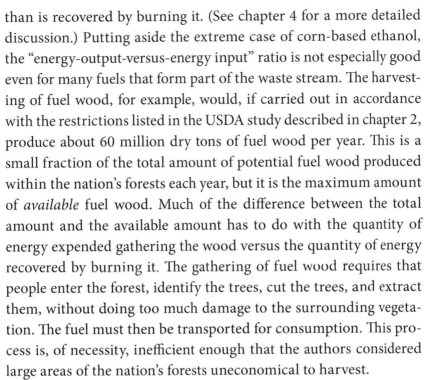

than is recovered by burning it. (See chapter 4 for a more detailed discussion.) Putting aside the extreme case of corn-based ethanol, the "energy-output-versus-energy input" ratio is not especially good even for many fuels that form part of the waste stream. The harvesting of fuel wood, for example, would, if carried out in accordance with the restrictions listed in the USDA study described in chapter 2, produce about 60 million dry tons of fuel wood per year. This is a small fraction of the total amount of potential fuel wood produced within the nation's forests each year, but it is the maximum amount of *available* fuel wood. Much of the difference between the total amount and the available amount has to do with the quantity of energy expended gathering the wood versus the quantity of energy recovered by burning it. The gathering of fuel wood requires that people enter the forest, identify the trees, cut the trees, and extract them, without doing too much damage to the surrounding vegetation. The fuel must then be transported for consumption. This process is, of necessity, inefficient enough that the authors considered large areas of the nation's forests uneconomical to harvest.

Another commonality shared by most biofuels—ethanol and biodiesel being two major exceptions—is that bioenergy projects are often not large and so fail to benefit from economies of scale. (Ten identical 100-kilowatt electrical generating units will, for example, generally cost more to build and operate than a single 1,000-kilowatt station of the same type, all other things being equal.) But the relatively low energy content means that it is uneconomical to transport these fuels long distances. As a consequence, power projects built to utilize biomass resources must be built near where the fuels are generated. They are, of necessity, local projects designed for specific local fuels.

Also, biofuels are seldom interchangeable. A power plant designed to burn woodchips, for example, may encounter difficulties burning grasses or biogas or many types of municipal waste. All of these factors serve to restrict the available fuel supply. Because these specially designed power plants cannot exceed the local supplies of fuel, they tend to be of small to moderate size—one more factor that contributes to higher overall costs.

EFFICIENT ENGINES

Internal combustion engines, diesel engines, and fuel-burning electrical generating stations all belong to a class of machines called heat engines. Heat engines power all planes, ships, and trucks, and virtually all automobiles—and heat engines generate more than 90 percent of the electricity produced in the United States. Heat engines are energy conversion devices: They convert thermal energy into the energy of motion or into electrical energy.

There is an enormous amount of engineering and scientific literature devoted to the study of heat engines. A great deal of 19th-century science was devoted to discovering the fundamental physical principles that govern the operation of heat engines, and a great deal of effort is still devoted to finding engines that operate as efficiently and reliably as possible.

The most important figure in the history of the scientific study of heat engines is the French scientist Nicolas-Léonard Sadi Carnot (1796–1832). Although his brief book, *Reflections on the Motive Power of Fire,* attracted little attention during his brief life, it contains a fully developed and essentially correct theory of heat engines. Incredibly, Carnot misunderstood the nature of heat—he believed it was an invisible fluid that obeyed the same basic physical principles as other fluids such as water. Nevertheless, he managed to discover the basic principles that govern the conversion of heat into work. In particular, Carnot discovered that there exists an upper bound on the amount of thermal energy that can be converted into work, and this upper bound is determined not by the fuel that is burned or by the device that burns it but by the operating temperature of the engine. (Recall that efficiency is the ratio of the amount of work produced by the engine to the total amount of thermal

Chevrolet Avalanche FlexFuel Vehicle. To make full use of the nation's ethanol production potential, consumers will have to buy large numbers of flex-fuel vehicles. *(General Motors Corporation)*

energy supplied to the engine.) In more modern notation Carnot discovered that

$$efficiency = \frac{T_o - T_e}{T_o} \qquad \text{(Equation 3.2)}$$

where T_e is the temperature of the environment and T_o is the operating temperature of the engine.

The first thing to notice about this equation is that this is the highest efficiency at which the engine can operate. It is an upper limit. While it is easy to build an engine that is less than optimally efficient, it is impossible to build an engine that exceeds this limit while maintaining the same operating temperature. Second, notice that if the operating temperature is identical to the temperature of the environment, the engine has zero efficiency, which is another way of saying that it can perform no work. And the larger the difference in

(continues on page 58)

Sustainability

"We, the representatives of the peoples of the world . . . reaffirm
our commitment to sustainable development."
    —*from the Johannesburg Declaration on Sustainable
    Development, September 2–4, 2002*

Sustainability is a term that is often introduced in discussions about
biofuels. Unfortunately, there is no single generally agreed-upon
definition of the term. One often-used definition is that today's generation should use resources in such a way that future generations are
able to meet their needs unimpeded. This is an ethical statement rather
than a scientific one, and it leaves open the question of how the criteria
of sustainability can be tested. A number of tests have, therefore, been
proposed for determining whether a particular practice is sustainable.
The tests differ, but the general idea is that resources should not be used
faster than they can, in some sense, be replaced, and wastes should not be
generated faster than they can, in some sense, be absorbed or recycled.

Applied to the production of biofuels, the concept of sustainable development is often used to mean that plant matter should be harvested
no faster than it can be replenished and that wastes, especially *greenhouse
gases,* should be generated no faster than they can be absorbed. For
example, with respect to the extraction of material from forests, sustainability means removing wood and other material no faster than it can be
grown on the same plot of land. If a plot of forest produces 20 cubic feet
of wood per acre (1.4 m³/ha) per year, for example, then this production
rate should also serve as an upper limit for the rate of wood removal.

Determining a sustainable rate for agricultural land is more difficult
since agricultural productivity has increased steadily since the middle of
the 20th century. Sustainable rates today are much higher than sustainable rates from the 1970s. Some of the reasons for increased productivity
could have been predicted decades ago—improved mechanization, for
example—but some of the reasons would probably not have been predictable several decades ago. *Biotechnology,* for example, has contributed

to higher sustainable yields per acre, and, in fact, approximately half of all acreage used to grow corn in the United States is now planted with genetically modified corn designed to improve productivity. These factors all contribute to a rate of sustainable yield that would have seemed impossibly high a generation ago.

There is, however, a more fundamental difficulty with applying the concept of sustainability: How can the needs of future generations be determined today? If it is impossible to determine the needs of the next generation, it becomes impossible to plan for them. Coal production, for example, is sometimes cited as an example of an unsustainable practice. To be sure, once burned, coal cannot be replaced, and so every ton of coal consumed today leaves future generations one ton of coal short. On the other hand, there is enough coal in the United States to maintain current levels of consumption for centuries, and it is difficult to argue convincingly that a century from now electricity will or will not be generated by burning coal. Consequently, it is difficult to argue that the consumption of coal today shortchanges future generations. If, during the next hundred years, coal-burning power plants are replaced by non-coal technologies, there would be enormous supplies of coal still in the ground at the time demand for coal collapses. Should this occur, any rate of coal usage today is "sustainable" in the sense that future generations will have abundant supplies of a mineral for which they have no use. (This example does not, of course, address objections to coal based on the pollution caused by burning it or the deaths that result from the mining of it.)

Despite the absence of an unambiguous and universal definition of sustainability and the absence of tests that provide clear answers about whether a particular practice is sustainable, ideas about sustainability of biofuel production can be helpful. It is useful to know (or at least it would be useful if one *could* know) whether the rates at which resources are used today can be extended into the foreseeable—that is, very short-term—future.

(continued from page 55)

temperature between the operating temperature of the engine and the environment, the more efficiently the engine can, in theory, operate. Because engineers cannot do anything about the temperature of the environment, they must design engines that have operating temperatures that are as high as possible. These high-temperature engines can, in theory, operate more efficiently than low-temperature engines in the sense that they can convert into work a larger percentage of the thermal energy obtained by burning the fuel. Efficiency is important because the more efficiently the engine operates the less fuel needs to be burned to produce the same amount of work. Reducing fuel consumption lowers costs and reduces the environmental impact of running the engine—all other things being equal. Conversely, anything that reduces the operating temperature of the engine—for example, moisture-laden fuel—can also reduce the efficiency of the engine and the amount of work performed per unit of fuel consumed.

As mentioned previously, most biofuels release less energy per unit volume when burned than the fossil fuels they most closely resemble. Consequently, engines that rely on biofuels can be expected, on average, to consume more fuel per unit of work performed. None of this is meant to dismiss the potential of biofuels, but it is one more limitation on the contribution that these fuels can make to the nation's energy supply. As will be seen in what follows, biofuels have a unique and important contribution to make to the energy supply of the United States and a number of other nations, but no large modern economy has learned to operate solely on biofuels. In fact, it remains an open question whether this is even possible.

PART II

Some Important Biofuel Technologies

Ethanol

Ethanol, also known as ethyl alcohol and grain alcohol, is a colorless, flammable liquid with a chemical formula that is written as either C_2H_6O or C_2H_5OH. The latter formula is used to emphasize the presence of the OH complex of atoms called hydroxyl. (Notice that both formulas indicate the same numbers of carbon, hydrogen, and oxygen atoms.) Ethanol has a boiling point of 172°F (77.8°C) and a lower heating value of 76,330 Btu per gallon at 60°F (21.28 MJ/l at 16°C). As a biofuel, it is used mainly in the transportation sector. Its most important contribution is as a fuel additive. It is also available at some service stations in richer gasoline-ethanol blends, such as E10, E85, and E100, which consist of 10 percent, 85 percent, and 100 percent ethanol, respectively. (E100 is something of a misnomer since ethanol is never sold pure; a toxin is always added to make the ethanol unfit for human consumption.)

A Colorado cornfield. Meeting the demand for ethanol with corn will require enormous amounts of prime agricultural land. *(USDA)*

Ethanol is the alcohol in alcoholic beverages, and methods of producing ethanol-containing beverages date back to antiquity. What is different with respect to the production of ethanol for use in the transportation fuels market is that huge volumes of almost pure ethanol are now manufactured solely to be burned. Production at this scale has required new ideas and new technologies to convert various ethanol feedstocks into fuel. But current technologies are not adequate to the task, and new and important changes in production methods are on the horizon, methods that may have far-reaching effects on the price and quantity of ethanol available.

This chapter begins by describing some characteristics of current large-scale ethanol production. It then describes an ongoing debate over ethanol's value as a transportation fuel. It concludes with a discussion of some characteristics of the ethanol market.

THE MECHANICS OF ETHANOL PRODUCTION

Ethanol can be produced from a variety of feedstocks, including corn, sugarcane, wheat, barley, potatoes, sorghum, sweet potatoes, and sugar beets. More broadly, ethanol can be made from any commodity that contains either starch or sugar. (In theory, grasses, trees, and the agricultural residue left on the fields after the crop has been harvested can also be used to produce ethanol, but the conversion process required for these feedstocks is not ready to be implemented commercially.) Today, only two feedstocks are in large-scale use: corn, the main feedstock in use in the United States, and sugarcane, the main feedstock in use in Brazil. Sugarcane is used to produce

Ethanol plant in Iowa (*Jim Parkin*)

ethanol in Brazil because in Brazil sugarcane is the least costly way to produce large amounts of sugar. Corn is used to produce ethanol in the United States because in the United States corn is the least costly way to produce large quantities of starch. These two nations produce approximately 70 percent of all the ethanol manufactured in the world today.

The idea behind ethanol production is simple enough. A feedstock is prepared—corn requires more preparation than sugarcane—and then yeast cells are added. The yeast converts some of the prepared feedstock into ethanol. The result of the process is a liquid-solid mixture, the liquid component of which contains water and ethanol. Finally, the ethanol is separated from the mixture in a multistep process to produce a liquid that is almost pure C_2H_6O. The challenge of the production process is to control the purity of the ethanol while producing billions of gallons of fuel each year at a cost that consumers can afford—or at least at a cost that governments can afford to subsidize. (In what follows, most of the emphasis is on the production of ethanol from corn. The United States produces 97 percent of its ethanol from corn. For more information on Brazil and the ethanol markets there, see the sidebar "Ethanol in Brazil," as well as the Further Resources section at the end of this volume.)

There have long been two main approaches to manufacturing ethanol fuel from corn. The technologies are called *dry milling* and *wet milling*. The distinction remains useful, although recent innovations in the dry milling process have begun to blur the differences in the methods. In 2007, in the United States, roughly 82 percent of all ethanol was produced in dry mills; the remaining 18 percent was produced in wet mills. Dry mills are often preferred to wet mills, because they are cheaper to build and so reduce investor risk. Their lower prices also enable small farmer cooperatives to construct their own plants.

The basic dry mill process begins by washing the corn, grinding it, and mixing it with water to form a mash to which enzymes are added to convert the starch present in the corn into glucose molecules. The mixture is then heated. Next, yeast is added, and it converts the glucose into roughly equal amounts of ethanol and carbon dioxide in a process called fermentation. The process stops when the ethanol has risen to about 12 to 18 percent of the mixture when measured by volume. At this point, the plant operators have a mixture of ethanol, water, and solids. The ethanol is separated in a two-step process. First, the mixture is heated in a process called distillation. Because the boiling point of ethanol at atmospheric pressure is roughly 40°F (20°C) lower than the boiling point of water, the ethanol evaporates first. The resulting ethanol vapor is recondensed apart from the mixture. The ethanol recovered during distillation still contains approximately 4 percent water, too much for it to be used as fuel. So the ethanol is then separated from the remaining water by a molecular sieve, a material that allows water molecules to pass through but not the larger ethanol molecules.

What is left behind after the ethanol has been separated is the water-soaked corn residue. The water is removed, and the result is a material called distillers dried grains and solubles (DDGS). DDGS is a valuable animal feed. Sometimes only some of the water is removed, producing a material called wet distiller's grains that can also be used as feed, but it must be consumed within a few days or it will spoil.

On average, the dry mill process produces 2.75 gallons (10.4 l) of ethanol and 17 pounds (7.7 kg) of DDGS for each bushel of corn produced. (A bushel of corn weighs 56 pounds [25 kg].) Sometimes ethanol mills are located near a feedlot so that the distiller's grains can be fed to cattle without drying. It is to the producer's advantage to avoid the evaporation process for DDGS, because drying requires energy, and energy costs money.

The wet mill process is more involved. The crucial step that distinguishes the two processes occurs near the beginning of the wet mill process, shortly after the corn has been washed. The corn is soaked in a special solution that allows the plant operator to

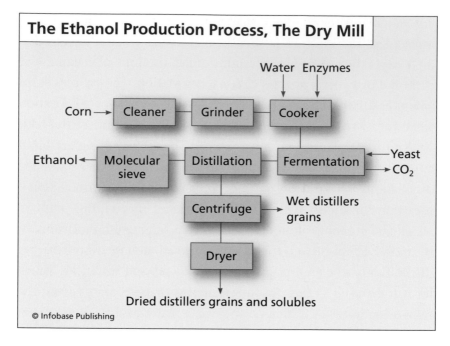

The Ethanol Production Process, The Dry Mill

Water   Enzymes

Corn → Cleaner — Grinder — Cooker

Ethanol ← Molecular sieve — Distillation — Fermentation ← Yeast → CO_2

Centrifuge → Wet distillers grains

Dryer

Dried distillers grains and solubles

© Infobase Publishing

Most ethanol is produced in dry mills, which, in addition to ethanol, produce large amounts of distiller's grains and carbon dioxide.

separate the starch, from which the ethanol is produced, from the rest of the corn kernel. By processing the corn more carefully, the plant operators are able to extract more coproducts, including corn oil, corn gluten meal, a poultry feed, and corn gluten feed, most of which is exported. The starch is then converted to glucose, which is fermented, and the ethanol, recovered. This part of the process is similar to that used in a dry mill and has already been described. The result of the wet mill process is (on average) 1.8 pounds (0.82 kg) of corn oil, 2.65 pounds (1.2 kg) of corn gluten meal, 13.5 pounds (6.13 kg) of corn gluten feed, and 2.65 gallons (10.3 l) of ethanol for each bushel of corn.

The amount and quality of coproducts obtained during the ethanol production process is crucial since their sale can mean the difference between profit and loss for plant operators. More recent innovations in the dry mill process now enable plant operators to begin to extract other coproducts such as corn oil, and so the distinction between the two processes is not as clear as it once was. Further refinements in the ethanol production process have also meant that the amount of energy expended in producing a gallon of ethanol has also continued to decrease from about 70,000 Btu per gallon (19 MJ/l) in 1970 to about 41,000 Btu per gallon (11 MJ/l) at today's most efficient dry mills. (Some ethanol producers have begun to switch from natural gas, the fuel most frequently used, to coal in order to save more on fuel costs.)

When the price of oil rises, the price of ethanol becomes more competitive relative to the price of gasoline. But even if gasoline becomes more expensive than ethanol, ethanol cannot replace gasoline. In the United States, the volumes of ethanol being produced are tiny compared to the amount of gasoline consumed each year. And because of ethanol's lower heating value—it is only two-thirds that of gasoline—cars can only drive two-thirds as far on a tank of ethanol as on a tank of gasoline. Consequently, it takes about 1.5 gallons of ethanol to replace a gallon of gasoline. If ethanol is to replace a significant amount of gasoline, it must be produced on an enormous scale. This is a big challenge, especially since efficient ethanol plants already convert more than 90 percent of the cornstarch into ethanol, so there is not much room for additional efficiencies in this area.

The simplest way to increase ethanol production is to plant more corn, but there are limits—limits on the amount of land that can be devoted to the production of transportation fuels instead of food and animal feed, and even tighter limits on the amount of land that should be devoted to transportation fuels. In any case, given the size of the gasoline market, no amount of U.S. farmland will be sufficient to satisfy even one-third of the nation's current demand for gasoline since current ethanol yields are in the range of 370 to 430 gallons of ethanol per acre of corn *per growing season,* and the

United States consumes almost 400 million gallons of gasoline *per day*—that is, in metric units, 3,460 to 4,020 liters per hectare per season versus 1.5 billion liters per day. If ethanol becomes competitive with gasoline on price, overproduction of ethanol will be impossible. How, then, can more ethanol be produced? What else can be done?

The first strategy is to increase the yield of corn per acre. United States farmers have been increasing their per-acre yields for many decades. By way of example, average yields during the early 1970s were, for the most part, well below 100 bushels per acre (6,200 kg/ha) and although there are always year-to-year fluctuations, yields exceeding 150 bushels per acre have become commonplace in recent years, a more than 50 percent increase over 1970s production levels. Continuing improvements in the per-acre yield of corn are expected into the foreseeable future as farmers continue to refine the methods by which they raise their crops and biotechnology companies continue to refine the genetics of corn. (Approximately 50 percent of the corn produced in the United States comes from genetically modified seeds, which were created to increase the hardiness of the plants and improve the yields obtained by the farmer.)

The biggest change in ethanol production will come if engineers are able to commercialize the production of ethanol from cellulosic biomass, a term that includes trees, leaves, grasses, and corn *stover*, the term used to denote the stalks and leaves of the corn plant, as well as many other sources of biomass. The process by which this is accomplished is called cellulosic biomass-to-ethanol or, sometimes lignocellulosic biomass-to-ethanol. ("Cellulosic biomass-to-ethanol," or *cellulosic ethanol* for short, is just the name of the process by which the ethanol is produced. The ethanol that results from the cellulosic ethanol process is chemically identical with the ethanol produced by more conventional means.)

From the point of view of the ethanol producer, there are two big differences between using corn and cellulosic biomass. First, corn kernels contain starch, which is easily converted into ethanol. There is little starch in cellulosic biomass. Instead, cellulose and certain related materials must be converted into ethanol. Cellulose is a material that contributes to the rigidity of plants, and cellulosic ethanol involves breaking down cellulose and related materials into simple sugars so that microorganisms can convert the sugars so obtained into ethanol.

The conversion of cellulosic biomass is significantly more difficult than the conventional process in which ethanol is produced from starch (or sugar), and despite a great deal of research some difficulties remain to be solved before the cellulosic ethanol process becomes commercially viable. Using present technology, the cost of producing cellulosic ethanol is about 50 percent higher than the cost of producing corn ethanol.

The second major difference between using corn kernels and, for example, corn stover as feedstock in an ethanol production process is that the conversion of stover yields little in the way of useful coproducts. Recall that in a corn kernel-to-ethanol production process, the production of coproducts can mean the difference between profit and loss. The production of ethanol from stover yields only plant residue and some methane. These can be used as fuel to provide process heat, but they are low-value products. It seems likely, therefore, that ethanol produced from stover will be priced higher than ethanol produced from corn kernels for the foreseeable future. The goal is to make the difference small enough to make cellulosic ethanol profitable.

The potential importance of stover arises from the fact that, when measured by weight, for every unit of corn harvested, one unit of stover is left in the field. Converting a substantial fraction of this stover into ethanol would greatly increase the per-acre yield of ethanol. But not all stover is available for use as a feedstock. In order to prevent erosion, about two-thirds of the stover would still have to be left in the field. This leaves one-third available for collection.

At 150 bushels of corn per acre, farmers achieve a yield of about 4.2 short tons of corn per acre (9.4 t/ha). This translates into about 1.4 oven-dry short tons of recoverable stover per acre (3.1 t/ha). With a maximum theoretical yield of about 107 gallons of pure ethanol

Ethanol in Brazil

The countries with the largest ethanol industries are the United States and Brazil. They produce ethanol at roughly the same rate per year, but when measured as a percentage of the national transportation fuels market or as a percentage of the national agricultural sector, the ethanol industry is many times bigger in Brazil than it is in the United States.

In Brazil, ethanol is produced from sugarcane. (Brazil has been a major sugar producer for much of its history and presently accounts for almost one-third of global sugarcane production.) The sugar-ethanol industry in Brazil accounts for 17 percent of agricultural output and two percent of the nation's gross domestic product. Somewhat more than half of all sugarcane grown in Brazil is used to produce ethanol. This represents an enormous commitment of agricultural resources. In 2007, roughly 15 million acres (6 million ha) were devoted to sugarcane production, and current plans involve expanding the amount of land devoted to sugarcane by 50 percent by the year 2012, at which point 3.6 percent of all arable Brazilian land would be used to grow sugarcane.

Brazil is also the world's pioneer in ethanol use, having developed a major internal market for ethanol in the 1970s. But it has encountered significant difficulties in creating a stable ethanol market. In the early 1980s, following the oil crises of the 1970s, approximately three-fourths of all new cars sold in Brazil ran on pure ethanol. When the price of oil collapsed in the 1980s so did sales of ethanol-powered cars. In 2000, sales of ethanol began to increase in step with the price of oil, but in 2003, there was an ethanol shortage. Flex-fueled vehicles, cars and light trucks that can run on pure ethanol or on any ethanol-gasoline blend, proved to be the solution to this type of instability. Today, close to 90 percent of new cars sold in Brazil are flex-fuel vehicles.

for each oven-dry ton of stover (446 l/t), farmers could, at current rates of production, increase ethanol production by as much as 150 gallons per acre (1,400 l/ha), an enormous jump in productivity. A more realistic ethanol yield of 75 percent of the maximum would

In 2007, almost all of the filling stations in Brazil offered motorists a choice between pure ethanol—what in the United States is called E100—and an ethanol-gasoline blend. Until 2006, this blend contained 25 percent ethanol, but ethanol shortages during that year led to price spikes, and the government responded by reducing the ethanol component of the ethanol-gasoline blend to 20 percent. In the spring of 2008, Brazil used a 24 percent ethanol-gasoline blend. Brazil hopes to increase ethanol production by approximately 62 percent by 2012 and is expanding sugarcane production and improving its energy infrastructure accordingly, including the construction of ethanol pipelines. Much of this additional ethanol will be for the export market. The United States market is seen as especially promising despite that nation's 54-cent-per-gallon tariff. (Brazil's ethanol industry is remarkably efficient. Brazilian sugarcane production costs are, for example, about two-thirds that of Mexico and less than half that of the United States, and it produces about 660 gallons of ethanol per acre (6,200 l/ha) so the high tariffs imposed by the government do not necessarily make exports to the U.S. market impossible, although a reduction in the tariff would lead to a surge in exports.)

Historically, the pattern of ethanol consumption in Brazil is complex, which serves to illustrate just how hard it is to replace petroleum in the transportation sector. But energy analysts now predict steady increases in Brazilian domestic ethanol consumption and Brazilian ethanol exports during the next ten years. Taken together, the high price of oil, flex-fuel technology, and increases in ethanol productivity mean that ethanol may soon become the principal transportation fuel for light cars and trucks in Brazil. This would be a historic change in energy consumption patterns.

still mean an extra 112 gallons per acre (1,050 l/ha), which would increase the ethanol yield by almost one-third. As strains of corn are developed that can be grown at higher densities, the yields from stover will increase as well. The barriers that prevent companies from producing cellulosic ethanol are economic and technical, but they are not insurmountable. Cellulosic ethanol is probably only a matter of time.

THE NET ENERGY BALANCE

One of the most controversial aspects of corn-based ethanol production involves the so-called *net energy balance,* a measure of the energy value of ethanol as a fuel. Arithmetically, the net energy balance is a fraction in which the numerator is the amount of energy obtained by burning a unit of ethanol, and the denominator is the amount of energy expended in producing the same unit of ethanol:

net energy balance = energy output / energy input

Unlike the energy output, which is determined by burning a sample of ethanol in a laboratory and measuring the amount of thermal energy produced, there is less agreement about the best way to compute the amount of energy expended in manufacturing a sample of ethanol—that is, the numerator in the net energy balance is well understood, but there is substantial disagreement about the *definition* of the denominator. (In what follows, only corn-based ethanol will be considered, because many of these ideas were first developed to understand the corn-based ethanol process. But the same ideas—although not the same numbers—would apply to other ethanol feedstocks and, more generally, other biofuels.)

Some aspects of what should be included in calculating the energy input are noncontroversial: For example, there is general agreement that one should count the amount of electricity used by the ethanol plant to produce its ethanol, and one should also count the gasoline, diesel fuel, or ethanol consumed by the farm machin-

Is more energy used to produce ethanol than is recovered by burning it? *(Tim Scearce)*

ery used to plant, harvest, and transport the corn that was used as ethanol feedstock. Another energy expenditure about which there is general agreement is the amount of natural gas burned during an ethanol production run. (Substantial amounts of natural gas are consumed making a batch of ethanol because distillation is an energy-intensive process.) There are also other energy losses that are less obvious but still significant and not especially controversial. Fertilizer for the corn, for example, is a necessary input when producing corn-based ethanol, and fertilizer production, although it occurs far from the farm, requires substantial amounts of energy, especially natural gas. Other input costs are less clear. Should one also count the energy used by the employees at the fertilizer factory to drive back and forth to work while they were producing the fertilizer? How far up the chain of production should one go in calculating the energy input?

Finding a reasonable measure of the total energy input required for the production of one unit of ethanol is important because one argument for producing ethanol, an argument that many find persuasive, is that the production of ethanol contributes to the energy supply. But if the net energy balance of corn-based ethanol is less than one—that is, if more energy is required to produce the ethanol than is recovered by burning it—then the production of ethanol represents an energy loss. To put it another way: If the net energy balance is less than one, the more ethanol that is manufactured, the greater the drain on the nation's energy supplies. Alternatively, if the net energy balance is approximately one, then the ethanol industry is "robbing Peter to pay Paul"—that is, it is expending valuable energy resources only to recover them in the form of ethanol, but not increasing the amount of available energy. A net balance that is significantly greater than one would mean that the production of ethanol contributes to the nation's energy supply. From a practical point of view, this is the most desirable of the three possibilities.

Much of the controversy about the value of corn-based ethanol was generated when a researcher named David Pimentel claimed that the net energy balance for ethanol is less than one. His analysis was criticized on two fronts.

> First, critics claimed that he used outdated information about ethanol production practices.

> Second, Pimentel's analysis included some of the energy used to manufacture the farm machinery and other less obvious "costs." The critics countered that if one is to include some of the energy used to manufacture the machinery, why not also include some of the energy required to manufacture the materials that were used to build the factories that were used to manufacture the farm machinery . . . and so on?

One step leads to the next with no end in sight, and is it not clear how much this type of analysis reveals about the value of ethanol as a transportation fuel. This second objection to Pimentel's analysis is less easily resolved, because what should be included depends on one's perceptions. There is no one right answer.

Several later analyses of the net energy balance—ones that used better data and different assumptions about what should be counted—yielded a range of values, all of which were somewhat greater than one. A more representative net energy value for these later studies is 1.34—that is, 34 percent more energy is obtained from burning ethanol than was used to produce it. Some studies that calculated significantly higher energy balances took into account the energy value of the coproducts, a practice that tends to obscure ethanol's contribution to the nation's energy supply.

None of these analyses indicate that the production of corn-based ethanol is an especially efficient use of valuable energy resources, but production continues. There are several reasons that resources continue to be devoted to the production of ethanol. Three of the more important are the following:

1. previous studies are not the final word on the subject of ethanol's net energy balance
2. production technology continues to improve (making the denominator in the energy balance smaller)
3. sufficiently large government subsidies can make ethanol production economically attractive even if the net energy balance is much less than one

Finally, it is important to keep in mind that the purpose of the net energy balance is to provide insight into what can be expected from the production of ethanol, but energy policy cannot be based on the value of a single fraction, informative as that fraction may be. There are other considerations besides the net energy balance in

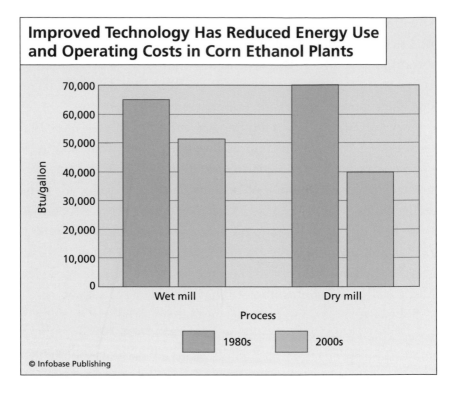

Improved Technology Has Reduced Energy Use and Operating Costs in Corn Ethanol Plants

© Infobase Publishing

A reduction in energy use means an increase in the net energy balance.

determining ethanol's value as a fuel, the most important of which is that ethanol is a transportation fuel. Much of ethanol's value lies in the fact that it is a partial alternative to oil, which is the essential transportation fuel. Most oil is consumed by the transportation sector, and the world's entire transportation sector—planes, cars, ships, trucks, trains, virtually everything that moves—depends on oil. In this sense, the transportation and oil sectors are mutually and completely dependent on one another. But the price and the supply of oil are volatile, which explains why all nations that are dependent on oil imports continually search for alternative transportation fuels. So far, alternatives have not been easy to find.

The other widely used primary energy sources—coal, natural gas, and uranium—are poorly suited for direct use in the transportation sector (although a tiny percentage of cars and light trucks have been modified to run on natural gas or on batteries). Presently, the only way to use these nonpetroleum primary energy sources in the transportation sector in a truly significant way is to use them indirectly in the production of ethanol. With this understanding, ethanol can be perceived as a method of storing some of the energy of coal, natural gas, and uranium for use in automobiles and light trucks. If one understands ethanol production from this point of view then the net energy balance hardly matters because ethanol is, then, an "energy carrier" rather than an energy source. It stores part of the energy of these other energy sources in a form that can be conveniently used in the transportation sector.

THE UNITED STATES ETHANOL MARKET

According to a 2007 study carried out at the University of Minnesota, if all of the corn grown in the United States were used to produce ethanol, it would replace only about 12 percent of the gasoline supply. The report concludes that "neither [ethanol nor biodiesel] can replace much petroleum without impacting food supplies." But the United States could never afford to convert its entire corn crop into ethanol. Corn is too important. Corn-based ethanol will, therefore, never really compete with gasoline as a national transportation fuel, because there will never be enough of it. This much is certain. But despite the tiny (compared to oil) ethanol market, ethanol has become an important component of the nation's fuel supply. To understand why, it helps to know more about gasoline, a complex fuel manufactured to satisfy a number of conflicting performance requirements.

One of the problems that early chemical engineers encountered in manufacturing gasoline was that once inside the engine's cylinders, gasoline would ignite at unpredictable times. The result was a phenomenon called engine knock, which occurred when gasoline ignited before the spark plug fired. Engine knock led to diminished automotive performance and premature engine wear. Lead was added to gasoline beginning in the early 1920s in order to better control the conditions under which the fuel would ignite. But the lead found its way out of the engine and into the atmosphere when the fuel was burned. As gasoline consumption grew, so did the levels of lead in the environment and in the blood of those exposed to auto exhaust. Lead is a poison, and elevated levels of lead in the body can cause nerve damage and learning disorders—especially among children. During the 1970s, lead levels in gasoline averaged 2–3 grams per gallon (0.5–0.8 g/l), resulting in the emission into the atmosphere of 200,000 short tons (180,000 metric tons) of lead by U.S. cars and light trucks each year.

During the late 1970s, the federal government began a program of reducing lead levels in gasoline with the goal of eventually eliminating lead altogether. Refiners responded to the mandate to reduce lead levels by substituting methyl tertiary-butyl ether (*MTBE*) for lead. The additive MTBE worked as intended: It prevented engine knock.

In 1990, Congress passed the Clean Air Act Amendments. In response the U.S. Environmental Protection Agency (EPA) began to study the problem of reducing levels of carbon monoxide, a poisonous gas, in auto exhaust, and in 1992, the EPA initiated the Winter Oxyfuel Program with that goal in mind. The cheapest and easiest way to accomplish the reduction of carbon monoxide was by adjusting levels of MTBE in gasoline. In 1995, the EPA initiated the Year-Round Reformulated Gasoline Program to reduce smog, which can also be accomplished by adjusting levels of MTBE to the gasoline. (By way of example, the optimal formulation for reducing carbon monoxide with MTBE is at a concentration of 15 percent by volume. For reducing smog formation, the optimal concentration of MTBE in gasoline is 11 percent by volume.)

But MTBE is not the only option for any of these applications. Ethanol blended with gasoline diminishes the possibility of engine knock. It can also be blended with gasoline to diminish carbon monoxide emissions and smog. For a long time, ethanol ran a distant second to MTBE in these applications, because in many ways ethanol is inferior to MTBE as an additive. MTBE could be mixed with gasoline at the refinery and distributed through the nation's enormous pipeline distribution system, but ethanol had to be mixed at the distribution terminal. It is ill-suited for distribution through the gasoline pipeline system. Moreover, MTBE-gasoline mixtures have a lower vapor pressure than an ethanol-gasoline mix. In practice, this means that the MTBE-gasoline mixtures evaporate more slowly and thereby contribute less to air pollution. Ethanol also has a lower energy content than MTBE (93,540 Btu/gal versus 76,330 Btu/gal [26 MJ/l versus 21 MJ/l] LHV). Finally, MTBE is cheaper than ethanol. There would have been little interest in ethanol except for the generous government subsidies paid for ethanol production. (By way of comparison, ethanol mixed at a concentration of 7.3 percent by volume reduces the emission of carbon monoxide to that obtained by using MTBE, and mixed at a level of 5.4 percent by volume, ethanol reduces smog formation in a way that is similar to that of MTBE.)

Today, there is little demand for MTBE. The problem is that MTBE, which has been identified as a potential carcinogen, diffuses easily through water—more easily than gasoline. When underground gasoline tanks leaked, MTBE quickly diffused through the groundwater—and the water became undrinkable. States began to ban the use of MTBE. This created an opportunity for ethanol producers, and so ethanol became important—not so much as a transportation fuel but as a fuel additive, a replacement for the fuel additive MTBE. (The term *fuel additive* implies that the volume of ethanol used is small relative to the volume of gasoline in which it is mixed, and that it is added for reasons other than its energy value, which was, initially, certainly the case with ethanol.) But because the volume of gasoline consumed in the United States is huge, the ethanol industry has been kept busy supplying what is, for it, a large amount of product.

What of using ethanol as fuel instead of as fuel additive? The Energy Policy Act of 2005 contains a provision that mandates increases in the amount of "renewable" transportation fuel used each year beginning in 2006. This mandate has contributed to the growth of the ethanol market, although ethanol's contribution as a transportation fuel, rather than as a fuel additive, remains small. This is not to say that the production of ethanol is negligible in the agricultural sector. It is not. If, as projected by the U.S. Department of Agriculture, ethanol replaces 7.5 percent of the gasoline used in the United States in 2017, almost one-third of the nation's corn crop will be required for ethanol production.

The big change, if any, in ethanol's status as a transportation fuel will occur when cellulosic ethanol becomes commercially feasible, at which time producers will be able to convert almost any plant feedstock into ethanol. Provided the price of oil is high enough (or government subsidies are large enough), volumes of ethanol can then be expected to rise sharply.

Biodiesel

Biodiesel is unusual among biofuels in that it is almost interchangeable with conventional diesel fuel. (In what follows, the word *petrodiesel* will be used to denote conventional diesel fuel.) By contrast, the performance characteristics of ethanol are very different from those of gasoline, and wood is very different from coal, but the differences between biodiesel and petrodiesel fuel are so small that under many conditions one fuel could be substituted for the other, and the user would be none the wiser.

Biodiesel is not a new discovery. Its proponents include the inventor of the diesel engine, the German engineer Rudolph Diesel (1858–1913). Inspired by the writings of Sadi Carnot, Diesel sought to invent an engine that would be optimally efficient, and his efforts led him to build the first diesel engine. Diesel was also an early proponent of biodiesel, and some of the earliest diesel engines used biodiesel fuel. In 1900, for example, a diesel engine that ran on

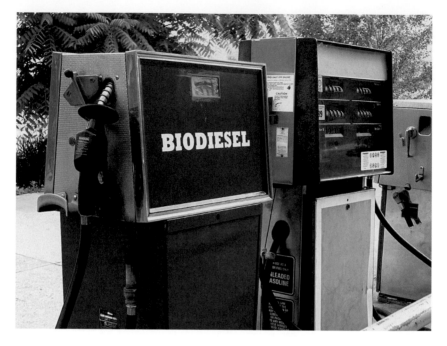

At current rates of production, biodiesel can only displace a very small percentage of the petrodiesel market. *(Logan Buell)*

peanut oil was famously displayed in Paris at the World's Fair. But despite its early start, biodiesel was quickly displaced by petrodiesel, which was cheaper and more plentiful. Interest in biodiesel did not revive until after the oil crisis of 1973. Today, the market in biodiesel is growing quickly. It is generally blended with petrodiesel fuel for use either in diesel engines or in the home heating market. (Conventional heating oil and petrodiesel fuel are interchangeable and are sometimes referred to collectively as "number 2 distillate.") This chapter describes what biodiesel is and how it is produced. It goes on to describe the economics of the biodiesel market.

SOME PROPERTIES OF BIODIESEL

To appreciate the properties of biodiesel, it helps to know something about the engine in which it is used. Diesel engines, unlike the internal combustion engines found in most automobiles, have no spark plugs. Instead, air is drawn into a cylinder, and a piston compresses the air inside the cylinder, usually between 14 and 24 times atmospheric pressure. (These are higher compression ratios than those used in internal combustion engines.) The rapid compression causes the temperature of the air inside the cylinder to increase rapidly, rising to about 1,000°F (540°C). Fuel is then injected into the cylinder. Mixed with the hot air, the fuel ignites almost instantaneously. The pressure inside the cylinder soars along with the temperature, and the piston is forced downward. It is the downward motion of the piston that performs the work.

Diesel fuel must be able to "auto-ignite" at the temperature and pressure conditions prevailing inside the engine—that is, the fuel must combust spontaneously upon injection into the cylinder. (By contrast, the fuel in internal combustion engines is ignited by a spark from the spark plug.) The design of diesel engines makes them more accommodating in terms of the fuels that they burn—Rudolph Diesel even experimented with an engine that ran on coal dust—and engineers have found that in addition to petrodiesel, a variety of oils can make acceptable fuels for ordinary unmodified commercially built diesel engines, provided that the oils are properly processed. Some of the plants whose seeds have been used as feedstock include canola (outside the United States this plant is better known as rapeseed), sunflower, peanut, mustard, soybean, coconut, and oil palm. Biodiesel has also been successfully made from used cooking oil collected from restaurants and food processing facilities.

But in order to use any of these feedstocks, the resulting oil must be processed, a fact discovered by researchers in the 1970s. The modern era of biodiesel research began during the 1970s in response to the oil crises of that decade. First, researchers in Austria at the Federal Institute of Agricultural Engineering (Bundesanstalt für Landtechnik), and later at the University of Idaho and in South Africa, began to independently tinker with the idea of producing biodiesel. They all began by simply substituting plant oil for petrodiesel. In Austria they operated a test engine using linseed oil, in Idaho they used safflower oil, and in South Africa they used sunflower oil. The result was the same in each case: At first, the engines ran smoothly, but the fuel left deposits that accumulated inside the cylinders. These deposits eventually destroyed the engines. The unprocessed oil was, they discovered, unsuitable for modern diesel engines. In retrospect, this is not surprising, because these engines had been designed to run exclusively on petrodiesel. The researchers were left with a choice: They could change the fuel, or they could change the design of the modern diesel engine.

Analyses and experimentation soon revealed that it was much easier and cheaper to change the fuel. Plant oils can be burned in commercially-built diesel engines without causing engine damage, provided the oils are chemically altered through a process called transesterification, which involves breaking up the larger oil molecules into smaller molecules. The result is biodiesel and a coproduct called glycerin (sometimes also called glycerol or glycerine).

Although alternative processing methods exist, most biodiesel refineries continue to use transesterification to produce biodiesel because it is relatively inexpensive and technologically simple. (Transesterification, which uses significant amounts of methanol, a toxin commonly used as a solvent, is simple enough that hobbyists sometimes use the process to manufacture their own biodiesel.) The energy demands of the biodiesel refining process are modest. Adding the energy used during refining to the energy expenditures required to produce common biodiesel feedstocks and dividing the result into the heating value of biodiesel yields a fairly high net energy balance. As a consequence, and in contrast to corn-based ethanol, there is, from an energy perspective, little controversy over the wisdom of biodiesel production. Accepted estimates for the net energy balance for biodiesel indicate that it is in excess of three—that is, more than three times as much energy is obtained by burning biodiesel as is used in producing it—and the balance

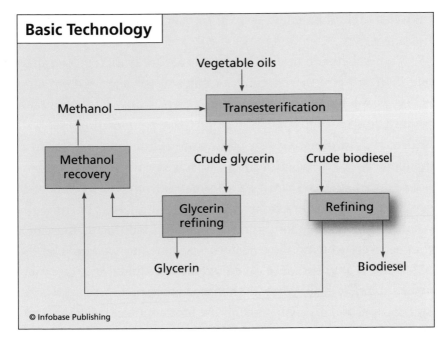

Transesterification is the reaction of vegetable oils and methanol to produce biodiesel and glycerin. (Notice that the methanol is recycled.)

is rising as production efficiencies improve. (See the section "Net Energy Balance" in chapter 4 for a more detailed discussion of how this statistic is calculated.)

As previously mentioned, when it is properly processed, biodiesel has performance characteristics that are similar, although not identical, to those of conventional petrodiesel fuel. Five characteristics of particular interest are the heating value of biodiesel, biodiesel emissions, its *cetane number* (defined later), its *pour point,* which is the temperature at which biodiesel ceases to flow, and its *lubricity,* which is a measure of biodiesel's value as a lubricant.

The heating value of biodiesel is comparable with that of petrodiesel. The lower heating values are 118,200 Btu per gallon (32.94 MJ/l) for biodiesel and 120,100 Btu per gallon (35.98 MJ/l) for petrodiesel—that is, the heating value of biodiesel is about 8 percent less than that of petrodiesel. (These values vary somewhat from sample to sample.)

Biodiesel greenhouse gas emissions are more difficult to calculate than those of petrodiesel. For petrodiesel, one needs to analyze the contents of the exhaust and the emissions associated with product transportation and refining. These computations are well understood and relatively straightforward, but for biodiesel such an approach would be misleading, especially with respect to greenhouse gas emissions. Growing the feedstock from which biodiesel is produced removes carbon dioxide from the air, just as burning the resulting fuel releases the carbon dioxide. It might seem, therefore, that between growing the feedstock and burning the biodiesel the net addition of greenhouse gases is zero, but while this is an oft-repeated claim, it is not correct. There are substantial greenhouse gas emissions associated with raising the biodiesel feedstock, harvesting it, transporting it, and manufacturing the fuel, and these should all be taken into account. Estimates for greenhouse gas emissions across the biodiesel production process, although they trend in the same general direction, are not uniform. These estimates are further clouded by uncertainties about current industry averages. That said, the U.S. Environmental Protection Agency estimates that across the "life cycle" of biodiesel—that is, greenhouse gas absorptions and emissions from production through consumption—net greenhouse gas emissions are reduced, relative to those of petrodiesel, by about two-thirds. (By way of comparison, corn-based ethanol achieves a reduction in greenhouse gas emissions of about one-fifth relative to gasoline.)

There are, of course, other emissions to consider besides greenhouse gas emissions. Most of these other measures of the impact of burning diesel fuel favor biodiesel over petrodiesel: Burning biodiesel reduces the emission of particulate matter by approximately 55 percent relative to petrodiesel; carbon monoxide emissions are reduced 45 percent, but nitrogen oxide emissions increase slightly (about 5 percent) when biodiesel is burned instead of petrodiesel.

(Nitrogen oxides are molecules produced during combustion, and they have a number of deleterious effects. Perhaps the most important are that they contribute to acid rain, and they contribute to the formation of ozone at ground level. Although ozone has environmental benefits when it forms high in the atmosphere, at ground level it can cause respiratory problems.)

It is important to keep in mind that these comparisons between pure biodiesel fuel and pure petrodiesel fuel represent laboratory tests. In fact, almost all biodiesel is consumed in petrodiesel-biodiesel blends in which the biodiesel component is 5 percent or less. In practice, therefore, the combustion of biodiesel results in very small changes in emissions because they are burned in fuel blends that consist almost entirely of petrodiesel. The same remark holds for all other properties considered in the chapter, except for lubricity, which is discussed later in this section. Differences between biodiesel and petrodiesel—even when they seem large in the abstract—generally make for very small differences in practice because of the highly dilute nature of the blends in which biodiesel is used.

Cetane number is a measure of the auto-ignition properties of a diesel engine fuel. All other things being equal, a higher cetane number indicates better fuel performance. For petrodiesel, cetane numbers vary from the low 40s in the United States to the middle 50s in Europe. For biodiesel, the cetane numbers vary from the high 40s to the high 50s.

The pour point for biodiesel, the temperature at which the fuel is no longer able to be pumped, presents a challenge for those using biodiesel in colder climates. The pour point for biodiesel fuel is higher than that of petrodiesel. This means that as temperatures fall, the fuel will thicken until it ceases to flow. It is not possible to give a single pour point for biodiesel, because the pour point depends somewhat on the feedstock used to produce a particular sample of biodiesel. Nonetheless, in all common types of biodiesel, pour points are higher than those for petrodiesel, which generally has a pour point of between -40°F and -30°F (-40°C to -34°C). Pour

Biodiesel clouding up. Biodiesel ceases to flow at much higher temperatures than petrodiesel. *(USDA ARS)*

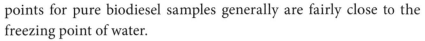

points for pure biodiesel samples generally are fairly close to the freezing point of water.

The final property of biodiesel discussed in this section is the lubricity of the fuel. Diesel engine fuel is also used as a source of lubrication for diesel engine fuel pumps. Biodiesel has somewhat better lubricity than low-sulfur petrodiesel fuel, long the standard fuel for diesel engines, but the difference was not large enough to affect consumption patterns. But low-sulfur petrodiesel has recently been phased out in favor of ultralow–sulfur petrodiesel, which has poor lubricity properties. By adding small amounts of biodiesel (1 to 5 percent by volume) to ultralow–sulfur fuel, the lubricity of the resulting blend is markedly improved. The demand for biodiesel is largely centered on the production of these low percentage blends.

THE BIODIESEL MARKET: DEMAND AND SUPPLY

Because some of the fundamental properties of biodiesel depend on the feedstock—the pour point and cetane number, for example—it might appear that there is not one biodiesel fuel but many. Differences in performance are, however, generally small. The real importance of the feedstock lies in its effect on costs, both economic and environmental. In the United States, most biodiesel is produced from soybeans. The reasons for this are largely historical.

In the early 1990s, there was only a small market for soybean oil in the United States, but there was a large market for other soybean products, particularly soybean protein for animal feed. As a consequence, there was a surplus of soybean oil. It was also at about this time that the first Gulf War caused a spike in energy prices. United States soybean farmers quickly identified the business opportunity created by simultaneously high petrodiesel prices and a large soybean oil surplus and organized to take advantage of it. To be clear, United States soybean farmers were not the first to create a biodiesel market—the European Union was the first to create a booming biodiesel market—but these early efforts by U.S. farmers were responsible for the U.S. market.

In 1992, farmers formed the National Soy Diesel Development Board, which, two years later, was renamed the National Biodiesel Board (NBB). Growth was slow at first—output was only one-half million gallons (1.9 million l) by 1999—but the NBB successfully lobbied for government subsidies, and beginning in fiscal year 2000, the USDA began a program called the Commodity Credit Corporation Bioenergy Program with the goal of encouraging biodiesel (and ethanol) production. The program had a positive effect on biodiesel production because it offered generous cash payments to producers. By 2006, the year the program was slated to expire, the market had grown to about 245 million gallons (927 million l), a 490-fold increase. In addition to the cash payments, a one dollar per gallon biodiesel tax credit remained in effect until 2008 under the Energy Policy Act of 2005. But these types of incentives also indicate problems with the biodiesel market. If biodiesel were truly competitive with petrodiesel, the incentives would not be necessary.

The choice of biodiesel feedstock is important for two reasons. First, biodiesel is expensive to produce, and the main cost is the feedstock. Currently, the cheapest biodiesel is produced from yellow grease, a material produced by food processors and restaurants. But it is not possible to run a nation's fleet of trucks on its supply of yellow grease. There is not enough of it. In the United States, estimates indicate that probably no more than 100 million gallons of biodiesel can be produced each year from yellow grease. For purposes of comparison, this is a small fraction of 1 percent of the U.S. diesel fuel market. Although soybeans remain the principal feedstock for the production of biodiesel in the United States, other nations use other feedstocks to supply their markets. No matter which feedstock is used, it remains true that at present no one knows how to make inexpensive biodiesel.

Despite the costs, governments worldwide continue to encourage biodiesel production and consumption through generous subsidies.

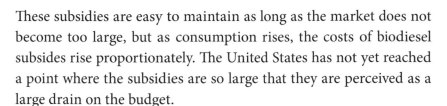

These subsidies are easy to maintain as long as the market does not become too large, but as consumption rises, the costs of biodiesel subsides rise proportionately. The United States has not yet reached a point where the subsidies are so large that they are perceived as a large drain on the budget.

Another way to compare feedstocks is on the basis of how much biodiesel is produced per acre of land cultivated. This is an especially important statistic if the policy goal is to replace significant amounts of petroleum with alternative fuels. For example, the U.S. market in number 2 distillate, the term used to collectively describe petrodiesel and home heating oil, is very large relative to the yearly production of the nation's biodiesel producers. In fact, biodiesel consumption remains at less than 1 percent of the total market. If, therefore, biodiesel production is to displace a large percentage of the number 2 distillate, it will require a very large shift in agricultural resources. One way to measure the shift is to calculate the amount of land needed to grow the necessary feedstock. The figures are daunting.

Soybeans yield only about 48 gallons of biodiesel per acre (450 l/ha). At that yield, there is not enough land in the United States to produce enough biodiesel to significantly change consumption patterns. This is not to suggest that soybeans should not be used to produce biodiesel. Soybeans were an important crop before the advent of the biodiesel market—soy protein is what made soybeans valuable originally, and soybeans remain important today for the same reason, as an excellent source of protein. Under these circumstances, the production of biodiesel as a soybean coproduct is efficient and profitable, but it is unrealistic to expect more than a small contribution to the diesel fuel market from soybean producers. The potential biodiesel market is simply too large to supply using soybean-based biodiesel.

The United States biodiesel market will, for the foreseeable future, be severely limited by the supply of biodiesel. In fact, it is mis-

(continues on page 94)

Biodiesel from Algae

One of the problems of biodiesel is that the production of large volumes of the fuel currently requires the use of large tracts of land. It is, in that sense, inefficient. Soybeans, for example, which comprise the main source of biodiesel fuel in the United States, produce only about 48 gallons of oil per acre (450 l/ha). This is a very low-density fuel source in the sense that if the soybean-based biodiesel obtained from a plot of land were spread uniformly across the plot on which the feedstock used in its production was grown, the oil would form a film about 0.002 inches (0.005 cm) thick. There is, however, at least in theory, an alternative that would completely change the biodiesel equation: the production of biodiesel from certain varieties of algae.

There are several advantages to using algae rather than higher plants for the production of biodiesel. First, the dry weight of some species of algae is as much as 60 percent oil. By contrast, it is sometimes claimed that some higher plants are 50 percent oil, but what is meant is that the *seeds* of the plant are 50 percent oil by weight. But the seeds are only a small part of the plant, and a great deal of energy is needed to produce the entire plant, which, in some cases, is grown solely for the purpose of harvesting the seed. That is certainly the case in soybean production, where only the bean is harvested. (The soybean plant minus its soybeans—often called the plant residue—is left on the field to prevent erosion and minimize soil depletion.) But with respect to algae, the 60 percent means exactly that—60 percent of the dry weight of the plant is in the form of molecules that can be converted to biodiesel.

Second, algae, because of their short generational time, are more susceptible to rapid genetic modification to improve plant characteristics and so improve yields.

Third, most conventional oil crops can be harvested just once or twice each year, but algae can be harvested much more rapidly. In fact, they can be harvested almost continuously.

Finally, because the algae would be grown inside enclosed structures, operators can manipulate the environment to optimize algal yields. Some early experiments have shown, for example, that diverting carbon dioxide–rich emissions from fossil fuel plants and sewage treatment plants to beds of algae accelerates the rate at which the algae grow. Even better, the algae convert the additional carbon dioxide into additional fuel. Such a system, if it could be brought to commercial development, would reduce greenhouse gas emissions while increasing energy availability. (It should be noted, however, that this scheme, if successful, would be more important to the algae producers than the power-plant operators, because the amount of carbon dioxide produced by coal- and natural gas–fired power plants far exceeds any conceivable capacity of agricultural operations to absorb more than a tiny fraction of it.)

Estimates of the amount of oil available per acre using algae as a feedstock vary widely. They reflect uncertainty about the productivity of a complex technology that is in its earliest stages of development, but one thing that all estimates have in common is that they are greater than the amounts of oil produced by more conventional feedstocks by roughly two orders of magnitude—that is, it is estimated that algae-based production methods will yield roughly 100 times as much oil per acre. Estimates range from 5,000 gallons per acre to 20,000 gallons per acre (47,000-190,000 l/ha). With such yields, algae cultivation could produce sufficient oil to replace substantial amounts of petroleum and do so without also monopolizing impossibly large amounts of agricultural land.

Intensive research into the following problems is under way:

1. identifying the most productive algal species
2. discovering the best ways to farm algae
3. creating efficient ways to harvest, dewater, and extract the oils

These are significant challenges, but they are the types of challenges that are amenable to intensive research. No nation is wealthy enough to maintain generous biofuel subsidies indefinitely while simultaneously increasing the size of its biofuel market. Germany, home of the world's most advanced biodiesel market, began to decrease subsidies in 2008 with the goal of treating biodiesel and petrodiesel equally by 2012. "Death by installments," is how Karin Ratzlaff of the Association of the German Biofuel Industry described the situation. She may be right. Biodiesel producers everywhere must become more efficient.

(continued on page 91)

leading to call biodiesel a transportation fuel, because that term mischaracterizes how most biodiesel is currently used. The most common biodiesel blends are B2 and B5. (Petrodiesel-biodiesel blends are identified with the letter "B" followed by a number that represents the percentage of biodiesel in the blend. B2, B5, and B20 are, respectively, blends consisting of 2 percent biodiesel and 98 percent petrodiesel, 5 percent biodiesel and 95 percent petrodiesel, and 20 percent biodiesel and 80 percent petrodiesel.) Because of the small volumes of biodiesel currently produced, only very modest amounts of B20 can be manufactured. But when sold as B2 or B5, biodiesel has little effect on the properties of diesel fuel because it is so diluted by petrodiesel. Only the lubricity of the fuel is significantly affected by the presence of the biodiesel in B2 and B5 blends. (As mentioned previously, ultralow–sulfur petrodiesel has poor lubricity properties, and so a fuel additive must be used to increase lubricity.) Even at 2 percent by volume, there is enough biodiesel in the fuel to lubricate those parts of the engine requiring lubrication from the fuel. At 2 (or even 5) percent, the other properties of the blend are similar enough to straight petrodiesel fuel that no other significant advantages are obtained by using the blend. It is no exaggeration, therefore, to say that at the present time, and for the foreseeable future, biodiesel really functions as a fuel additive rather than as a transportation fuel. It is valuable, but it is hardly a realistic alternative to petrodiesel.

In Europe, the biodiesel market is more fully developed than in the United States. In 2006, European sales of diesel-powered passenger cars exceeded those of gasoline-powered passenger cars for the first time. In France, in particular, almost three-fourths of all cars sold during 2006 were diesel-powered. This is, therefore, a much broader market for diesel fuel than exists in the United States. Furthermore, this market has been developed in a way that encouraged the rapid growth of biodiesel feedstock, while at the same time minimizing any immediate effects on food prices. Farmers in the European Union

(EU), for example, are bound by the EU's Common Agriculture Policy. In particular, the policy prohibits farmers from growing food or feed crops on 10 percent of their arable land. Farmers are, however, permitted to grow "industrial" crops on this land. This policy reduces the initial impact on food and feed prices caused by shifting production from food and feed to energy crops, because, at least on the set-aside land, energy crops did not initially displace any food or feed crops. Coupled with substantial production subsidies, the result was an initial period of rapid growth in energy crops, especially canola, the biodiesel feedstock of choice. Oil yields with canola are substantially higher than those for soybeans, roughly 130 gallons per acre (1,200 l/ha), more than twice that attained with soybeans.

The use of canola as a biodiesel feedstock is an efficient way to use agricultural land in temperate climes—more efficient, at least, than other feedstocks grown in temperate regions. But in tropical regions, producers prefer coconut and oil palm for feedstocks. The difference is significant because coconut and oil palm plants have yields that are much higher than their temperate counterparts. Using coconut, growers can obtain yields of about 290 gallons of oil per acre (2,700 l/ha), and oil palm can yield 635 gallons of oil per acre (5,930 l/ha), which is an enormous improvement in per-acre yield over what can be achieved in temperate regions.

As worldwide biodiesel markets grow, developing nations have begun to aggressively increase biodiesel production capacity, and most of their production is for export. The boom in tropical energy farms has generated controversy on three fronts. First, some claim that large energy crop farming operations are displacing—or at least have the potential to displace—subsistence farmers and small-scale commercial farmers. These claims have attracted a lot of attention but, at least so far, little evidence has been presented to support them. The claims may or may not be true. Without research there is no way of knowing.

A second objection to the establishment of some large-scale energy farms has to do with the conversion of forests to farmland.

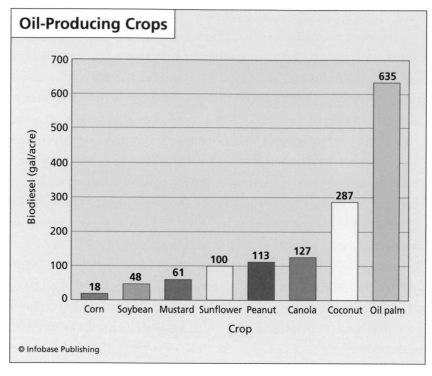

Biodiesel yields in gallons per acre for a variety of crops

This objection was raised by Achim Steiner of the United Nations Environment Programme, who, in November 2007, expressed concerns about the use of fire to clear Indonesian forests in order to create farmland to grow palm oil crops. The burning of these tropical forests releases large amounts of carbon dioxide, so much carbon dioxide that it may not be possible to recoup the emissions by the production of biofuels from these same lands. Mr. Steiner sought to give his argument an economic emphasis by expressing concern that once consumers become aware of the environmental havoc caused by the forest-clearing operations, they may refuse to buy biodiesel produced from these regions.

The forest-clearing operations have, however, continued unabated. After all, consumers have demonstrated little interest in the conditions under which petrodiesel is produced, and the deleterious impact of oil production on some developing nations as well as the broader environment is well documented. Given that the motivation for the establishment of these tropical energy farms in the heart of large forests *is* economic and not environmental, and Mr. Steiner's objections are fundamentally environmental and not economic, it is doubtful his remarks will be heard where it matters. Regardless of the outcome of the Indonesian controversy, it is clear that developed nations must increasingly look to international markets to satisfy increased domestic demand for biodiesel. For no matter which feedstock is used (see the sidebar "Biodiesel from Algae" for the one possible exception to this statement), producing enough biodiesel to replace a large fraction of petrodiesel consumption will require staggering amounts of land, amounts that often exceed what a large developed country can dedicate to its own fuel production.

The final objection to large-scale biodiesel production—and the same arguments apply to ethanol—rests with the surging price of grains. For the poor, an increase in price is the same as a decrease in supply. Sharp increases in grain prices have occurred as biofuel production has increased sharply, but the connection between the two is not entirely clear. Fossil fuels have also increased in price simultaneously with increased biofuel production, and fossil fuel prices have also caused an increase in grain prices. Disentangling the effects of these two factors on grain prices is not easy. And changing diets have also affected demand as the standard of living in many countries has improved. When hundreds of millions of people change from eating one meal per day to eating two, the effect on demand is substantial. And as the newly affluent eat more meat, the demand for grain must increase sharply as well. (It takes roughly two units of grain to produce one unit of beef and four units of grain to produce one unit of pork.) Biofuel production must, of course, contribute to grain shortages, but the principal factors may lie elsewhere. Not enough research has been done in this area to know for sure.

There are serious issues associated with converting land from food production or from forest to energy production, and these issues will come increasingly to the fore as worldwide biodiesel production continues to increase. There is no avoiding the discussion. To balance the supply of biofuels with the potential demand will involve creating an enormous new agricultural industry. The characteristics of that industry have yet to be determined.

6

Electricity Production

The main applications for heat engines are in the transportation sector and in the power-generation sector. The transportation sector was discussed in chapters 4 and 5, and the power-generation sector is the subject of this chapter.

Biofuels have not made the same contribution to the power-generation sector that they have made to the transportation sector. In part, this reflects the fact that fuel "competition" is stiffer in the power sector, where multiple power-generation technologies and multiple fuels are already in wide use. Heat engines produce the most electricity, and the principal fuels are coal, uranium, and natural gas. Other methods of generating power—methods that do not involve the conversion of thermal energy to electrical energy— include wind power, hydroelectric power, photovoltaics, and wave power. Hydroelectric power has long been a mainstay of the power-generation sector, and the contribution made by other nonthermal

Wood-fired, 65-MW power plant in Hovinsaari, Finland *(Kotkan Energy)*

generating technologies, while small, is growing rapidly. Geothermal power, which is currently limited by the lack of suitable sites, may also be poised for a period of rapid growth as new technologies are developed. In order for a biofuel to become a significant source of energy in the power sector, it would have to satisfy a so far unmet need or outperform an energy source already in place. This has proven to be difficult.

Broadly speaking, the consumption of biofuels for the production of electricity can occur in one of two ways. Biofuels can serve as the sole source of power for the production of electricity (at least some of the time), or biofuels can be blended with fossil fuels in a process called *cofiring*. Both of these technologies are examined in this chapter, but first it is necessary to review some basic facts about power generation and the economics of power-plant operation.

ELECTRICITY PRODUCTION AND THE ELECTRICITY MARKETS

All power plants, whether or not they are heat engines, are energy conversion devices, and with a few exceptions—and these exceptions generate very little total power—all power plants convert the straight-line motion of a liquid or a gas into rotary motion; the rotary motion is used to turn the shaft of a generator, and the generator produces the electricity. No matter how sophisticated or simple the design, spinning the shaft of a generator remains the main function of most power plants.

The method by which the generator is spun is most easily observed in hydroelectric plants, where the straight-line motion of water is converted into rotary motion by a device called a *turbine,* a piece of machinery that is conceptually similar to a waterwheel but much more sophisticated in design. (The generator is connected to the turbine by a shaft, and as the water spins the turbine, the generator produces electricity.) Heat engines also depend upon turbines. To spin its turbine, a heat engine is equipped with something called a *working fluid.* Often, but not always, the working fluid for a heat engine is water, in which case the turbine is driven by the power of expanding steam.

Usually, in a coal-, biofuel-, or natural gas–fired power plant, the heat generated by combustion is transferred from the combustion chamber through a radiatorlike device, called a heat exchanger, to a supply of circulating water. The heat exchanger permits the transfer of heat but not mass from the combustion chamber to the liquid water. The water, which is sealed within a system of pipes, absorbs enough heat to turn into steam, but because the volume occupied by the water is fixed by the volume of the pipes, the steam is formed and maintained at a very high pressure. Once produced, the steam is pumped through pipes to the steam turbine, where it is released through a valve. The steam, which passes through the valve at high velocity, strikes the turbine blades, causing the turbine to spin. Once past the turbine, the steam is recovered and cooled. As it cools, the steam condenses and reverts to liquid water. The liquid water is pumped back to the heat exchanger, and the cycle begins again.

Whether a plant, once built, actually produces power depends on whether it is profitable to operate. For the majority of U.S. power plants, the economic conditions under which they now operate are very different from those envisioned by those who designed them and financed their construction. For most of the history of the electric-power industry in the United States, electric utilities were local monopolies that built the power plants that they needed, used their own facilities to produce the power that they sold, and sold the power they produced to the individuals and companies living in their market, a market for which they were the sole suppliers. They were, in short, monopolies, and monopolies, which have sometimes been associated with high prices and bad service, have never had a very good reputation in the United States. Consequently, governments—local, state, and national—sought to limit the power of the electric utility monopolies in various ways. Sometimes municipalities formed their own electric utilities in order to be sure that the essential service of power generation was performed in a way that served the common good. More frequently, governmental regulatory bodies were established to monitor the behavior of the privately owned utilities and enforce compliance with various regulations—all this to ensure that the public good was served while simultaneously guaranteeing investors a reasonable profit, thereby ensuring the financial health of the companies that produced the electricity.

The system of strict governmental regulation worked reasonably well for many decades, but as with any system it had its flaws. During the last two decades of the 20th century, certain problems arose within the industry that were not easily addressed by regulation. One

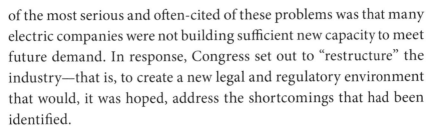

of the most serious and often-cited of these problems was that many electric companies were not building sufficient new capacity to meet future demand. In response, Congress set out to "restructure" the industry—that is, to create a new legal and regulatory environment that would, it was hoped, address the shortcomings that had been identified.

In 1996, in response to new federal legislation, the Federal Energy Regulatory Commission (FERC) issued orders 888 and 889. These orders effectively restructured the electric power industry. The following is one of the most important passages from these orders:

> [Order 888 requires] all public utilities that own, control or operate facilities used for transmitting electric energy in interstate commerce to have on file open access non-discriminatory transmission tariffs that contain minimum terms and conditions of non-discriminating services . . .

Whereas the old utilities had performed several functions, among them producing power, transmitting power over high-voltage networks, and operating local power-distribution systems, the restructured system "unbundled" these functions so that they were often performed by different entities. The idea was that local utilities would buy power from multiple competing power producers and that purchasing decisions would be made on the basis of price alone, thereby encouraging competition among producers. Central to this concept is the idea that high-voltage transmission lines be operated by a neutral authority—that is, that the rates or "tariffs" charged to power producers would be "non-discriminatory"—in the sense that any company capable of landing a contract with a local utility had an equal opportunity to send the power from its generating station to the utility.

Unbundling was supposed to encourage competition between power producers. Each power producer, it was hoped, would try to underbid the others, offering lower-cost electricity to the local utility. Those producers that could innovate and produce power more cheaply would prosper, and open competition among many different producers would lead to increased investment and lower prices. Meanwhile, power producers that could not price their electricity competitively would not make the necessary sales, and their plants would sit idly and unprofitably. That, at least, was the theory.

To see the ways in which power producers compete, it is important to keep in mind that the demand for electricity is only partly predictable. Local utilities are able to accurately plan for the power demands made by hospitals and other 24-hour public services, for the power consumed by streetlights, for example, as well as some manufacturing concerns, and some of the demands of homes, schools, and businesses. These power requirements are highly predictable. Taken together they form what is called the *base load,* and local utilities generally sign long-term contracts with power producers to supply base load power. Nuclear plants, coal plants, and some hydroelectric power producers are especially well-suited to the task of providing base load power because their performance is highly reliable, and they can operate for long periods at relatively low cost. While there is some competition to provide base load power, it is limited in extent.

But in addition to base load power, there is also *peak power,* which involves providing electricity for the additional demands made upon the system above the base load requirements—demands that are not entirely predictable. By way of example, everyone knows that there will be hot days during the summer when power demand will soar due to the use of air conditioning equipment, but no one knows far in advance precisely when those days will occur. Peak power demands are satisfied by short-term contracts between producers and utilities. Some peak power contracts are made the day before, and some contracts are made only minutes before the power is needed.

In New England, the power market in which one of the biofuel-fired power plants considered in this section is located, the system

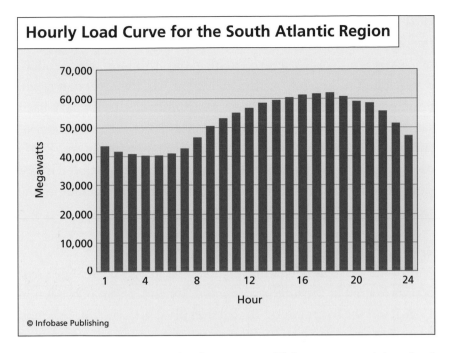

Hourly load curve for the South Atlantic region. While most power is base load power, there is significant variation in demand over a 24-hour time frame. *(Source: North American Electric Reliability Council)*

works as follows: A nonprofit organization called ISO New England—the ISO stands for "independent system operator"—continuously monitors the grid in order to determine real-time demand and then acts to ensure that demand and supply are continually in balance. Independent power producers enter their bids to supply the necessary power according to a specified procedure, and as demand climbs above the base-load requirements, ISO New England begins to accept bids—lowest first—and continues to climb the "cost ladder" until the demand for power is satisfied. Later, as demand begins to fall, ISO New England begins to issue orders for producers to shut down. The highest cost producers are shut down first. ISO New England, as a neutral authority, has the responsibility of managing the day-to-day operation of the New England power market.

(It is also responsible, for example, for routing the power through the high-voltage lines so that the power can arrive at its destination efficiently.) In the restructured market, bids from low-cost producers are often successful, but the higher the bid, the less often it will be accepted. Unless demand is very high, it will already have been satisfied by the lower cost producers before it becomes necessary to call upon the higher cost producers to make a contribution. Other regional electricity markets in the United States are operated in the same general way, although the details vary from one region to the next.

BIOFUEL-FIRED POWER PLANTS

One of the best-known biofuel plants is the Joseph C. McNeil Generating Station in Burlington, Vermont. When it began operation in 1984, it was the largest wood-fired power plant in the world, although its output was a modest 50 megawatts (MW). (By way of contrast, the Vermont Yankee Nuclear Power Plant, located in Vernon, and which began operation in 1972, has an output somewhat in excess of 500 MW—small for a commercial nuclear plant but an order of magnitude larger than McNeil.) The original design of the McNeil Generating Station was in many ways similar to a small conventional coal-fired plant, the principal differences arising mainly from the choice of fuel. In fact, when McNeil was first constructed it was built so that it could be converted to a pulverized-coal–fired power plant, should the need arise. It has never burned coal.

McNeil has been modified both for economic and demonstration purposes over the years. In 1989, when natural gas prices were relatively low, it was modified to burn natural gas. During the late 1990s, the price of natural gas rose sharply. As a result, it no longer makes economic sense for the McNeil plant to burn natural gas. As of this writing, it costs about twice as much for McNeil to produce a kilowatt-hour of electricity using natural gas as it does with wood and three times as much to produce a kilowatt-hour of electricity using oil, which it can also burn, as it costs with wood. Natural gas and oil are now used only during start-up. Because the price of these fossil fuels will, over the long term, only increase, and because the cost of wood is fairly stable, the competitive advantage of using wood relative to these other fuels can be expected to increase.

During the late 1990s, McNeil was also modified to demonstrate the feasibility of a *gasifier* unit. Gasifiers convert solid fuels into gaseous fuels, and in theory, a gasifier offers the potential to remove pollutants from the biomass before it is burned. The gas that is produced is, in that sense, the cleanest part of the original fuel. Gasifiers, however, are expensive. The gasifier that was built at McNeil was a demonstration unit, and it has not been used since 2001, when the owner of the unit went into bankruptcy. Twenty-three years after it began operation, the McNeil plant operates with essentially the same equipment and in essentially the same way as its designers originally intended.

McNeil depends on a traveling grate stoker (also known as a chain grate stoker), a long conveyor belt–like device that moves the wood chips from storage through the combustion chamber, and then deposits the ashes for removal. The traveling grate moves slowly through the chamber, and as it moves, its payload (the wood chips) burn and produce the heat needed to convert water (the working fluid) into steam. By the time the wood chips have reached the end of the conveyor belt, there is nothing left but ash. This is the same type of system that is used in some small coal-fired units.

Relative to common fossil fuels, wood does not have a very high heating value, and the McNeil generating station is not as efficient as a fossil fuel–fired plant in the sense that it does not convert as much of its thermal energy into electrical energy as a similarly sized, well-maintained fossil-fuel plant. Its thermal efficiency is about 25 percent compared to a conventional coal-fired plant, which converts about one-third of its thermal energy into electrical energy. But because the amount of CO_2 released by McNeil during the combustion pro-

The wood-fired, 50-MW McNeil Generating Station in Burlington, Vermont (*McNeil Generating Station*)

cess is reabsorbed by trees as they grow, McNeil contributes far less in the way of greenhouse gases. (Other emissions, notably sulfur dioxide, a key contributor to acid rain, and mercury, a toxic metal, are serious problems for coal-fired power plants but are absent from the emissions produced by McNeil, because its fuel contains neither sulfur nor mercury.)

It may seem as if the amount of carbon dioxide emitted during power production and the amount absorbed by trees as they "grow into fuel" balance each other out, but as with other biofuels, there is a certain amount of energy expended in collecting, preparing, and transporting the fuel, and these emissions must also be taken into account. In the case of wood fuel, however, the difference is small, and McNeil is nearly carbon neutral—that is, the amount of carbon emitted divided by the amount of carbon absorbed is almost equal to 1. To maintain its maximum power output, the McNeil plant consumes about 76 short tons (69 t) of wood chips per hour.

Approximately 70 percent of the wood chips that the McNeil Station burns are obtained from two sources: harvest residues, by which is meant branches and unusable tree trunks left behind by commercial logging operations, and low-quality trees, a term reserved for trees with no other commercial value than as fuel. Many trees, of course, are not of low quality. They have commercial value as lumber or as furniture, and it would be much too expensive to purchase these more valuable trees simply to burn them. Other trees, of low quality or not, are located in parks or ecologically sensitive areas and are unavailable for harvesting. As a consequence, the number of low-quality trees available for sustainable harvest is much smaller than might first appear. The U.S. Forest Service estimates that the amount of low-quality wood available throughout northern Vermont is about twice what McNeil needs to operate if it were operated full time at about 85 percent of its maximum output. (Keep in mind that northern Vermont is heavily forested.) There is, therefore, plenty of wood in the region for one 50-MW wood-fired generating station, but according to the U.S. Forest Service, the entire northern half of the state could not produce sufficient resources for two more such plants, and supplies would be tight for even one more McNeil-type woodchip–fired power plant if it were located near McNeil.

Of the remaining 30 percent of the wood consumed by the plant, about 25 percent is purchased from area sawmills in the form of sawdust, bark, and other waste products. The remaining wood (about 5 percent of the total) is municipal waste.

In practice, McNeil obtains its fuel from whichever source is cheapest and has ordered wood from sources as far as a few hundred miles away. Maintaining the flow of wood fuel is a continual challenge. When too much wood accumulates at the plant, it begins to rot and smell, and the plant's neighbors begin to protest. (McNeil is located within the Burlington city limits.) Too little wood on site, however, and the plant would not be able to meet its contractual obligations. Because contracts are often confirmed the evening prior to the day the power is needed, and because it can take up to eight hours to bring the plant to operating temperature, maintaining the right amount of fuel on site will remain a challenge.

As a commercial enterprise, McNeil competes with other New England power producers in the short-term, peak-power market. Each producer places its bid according to the procedure described in the preceding section and, provided the bid is low enough, sells its power to utilities in the New England market. During the summer and winter, the seasons during which demand for electricity in New England is highest, McNeil produces power almost continuously. During the spring and fall when average demand falls, it produces power intermittently, starting and stopping according to the demands of the market. In 2007, it had a *capacity* factor of about 70 percent—that is, it produced about 70 percent of the power it could have produced if it had operated at full power the entire year. By way of contrast, nuclear plants generally operate in the vicinity of 90 percent capacity, and wind turbines generally operate in the 20s.

McNeil has had some challenging years, times when its owners questioned whether it was worth operating. Today, it is a small but viable power producer in a highly competitive market. The plant has not changed. Rather, the economics of producing power are radically different from the conditions that prevailed when it first opened, and by persevering it has prevailed as a biofuel-only power plant. Its contribution to the market is, however, small, as are those of other biofuel-fired power plants.

Power producers can use biofuels without limiting themselves to relatively small plants by cofiring biofuels together with fossil fuels. The term "cofiring" means that a biofuel is simultaneously burned with a fossil fuel—wood with coal, for example. Cofiring at a coal-fired power plant may make economic sense, provided the price of coal is relatively high and local supplies of biomass are abundant and inexpensive. Cofiring can be worthwhile for the biomass provider if the alternative is disposing of the biomass at a local landfill, where disposal costs are high. Absent any other market, it may make good economic sense to simply give the biomass away. But even when the biomass is free, cofiring may not always be worthwhile for the power-plant operator. (See the sidebar "A Case Study: Power Generation versus the Environment" on the following page.)

To appreciate how cofiring works at utility-scale coal-fired power plants, it is important to keep in mind two characteristics of these power plants. First, they are often enormous in order to benefit from economies of scale. To meet the demand for electricity, they must burn many hundreds of tons of coal each hour. Under these circumstances, coal cannot be entirely displaced with wood. There would never be enough wood. Nor can wood be substituted for coal on a ton-per-ton basis. The value of a fuel is measured by its heating value, not its weight. Wood generally has a lower heating value than coal. Green wood, for example, may have a heating value that is only about 60 percent that of coal. When measured on a per-mass basis, therefore, it may require about 1.7 times more wood

(continues on page 114)

A Case Study: Power Generation versus the Environment

In 2003, the U.S. Department of Energy's Office of Energy Efficiency and Renewable Energy (EERE) released a study detailing its investigation into the possibility of using poultry litter, a combination of animal waste and animal bedding, to produce electricity. The EERE had been hoping for a demonstration of biomass power generation technology at the Reid Power Plant, a conventional coal-fired facility located in Sebree, Kentucky, and owned by Western Kentucky Energy Corporation. (This facility had been modified to also burn natural gas in order to reduce nitrogen oxide emissions during the warmer months of the year when these emissions are most problematic.)

Southwest Kentucky is home to a number of commercial poultry farms. At the time of the study, more than 500 poultry farms were situated within a 50-mile (80-km) radius of the Reid plant, and these businesses generated more than 75,000 short tons (68,000 t) of poultry litter each year. Disposing of so much litter is, from an environmental viewpoint, a difficult problem. Turning this waste into electricity would serve two important functions: First, the scheme would enable the farmers to solve a waste-disposal issue, and second, Western Kentucky Energy would get access to a renewable source of fuel to generate electricity. It seemed as if both parties would benefit.

The energy conversion scheme required the installation of a gasifier, a device that would convert the litter into a combustible gas, so it would be less polluting and easier to burn. The gasifier was designed to handle 8.4 short tons (7.6 t) of waste per hour and produce gas with a heating value in the range of 85 to 140 Btu per cubic foot (3,200–5,200 kJ/m^3). (The variation in heating value is the result of variation in the expected composition of the litter.) By way of comparison, natural gas has a heating value of about 930 Btu per cubic foot (34.6 kJ/m^3), a value that is between six and 11 times higher than that from the renewable source. For this reason the gas produced from the litter is classified as a "low Btu" fuel. But cofiring poultry litter–derived fuel with coal would reduce emission levels of

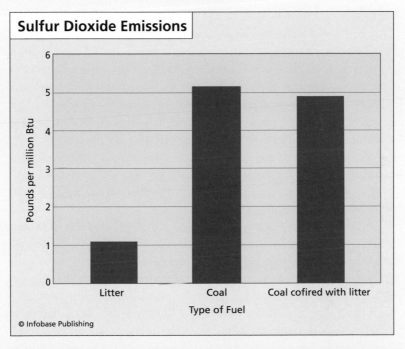

Burning straight litter results in a sharp drop in sulfur dioxide levels, but litter supplies are insufficient to power the plant. Compared to straight coal, cofiring results in a marginal reduction in sulfur dioxide levels, another indication that local farmers would benefit more from cofiring litter than the power plant operator. (*Source*: Gasification Based Biomass Co-Firing, Phase I: Final Report. *Prepared by Nexant for NREL*)

certain key pollutants, especially sulfur dioxide, relative to the emissions produced by burning coal alone.

The Reid plant is a substantial facility. One way of appreciating its energy requirements is that it burns enough fuel to produce 690,000 pounds (313,000 kg) of steam per hour at 955°F (513°C) at 89.4 atmospheres. There are many similar coal-fired units located around the nation, a fact that reflects

(continues)

(continued)

the enormous demand for electricity in the United States. But if the Reid plant is ordinary by the standards of commercial coal-fired generating units, its fuel requirements cannot possibly be met by the low Btu fuel manufactured by the gasifier. Engineers estimated that under the best of conditions, only 8 to 10 percent of the requirements of the Reid plant could be met with the poultry litter–based fuel. Consequently, whatever advantages the poultry litter fuel could provide—and there were several—the difference made by cofiring this fuel would, from the power plant operator's point of view, be small, because even with the gasifier working at full capacity at least 90 percent of the plant's thermal energy would still be derived from fossil fuels.

In the end, after the study had been completed and the costs and benefits of the scheme had been evaluated, Western Kentucky Energy decided not to go ahead with the installation of the gasifier, because it was, in their estimation, uneconomic to do so. This illustrates the problem with many renewable energy schemes: Although the poultry litter–to-electricity proposal was described as an energy-producing scheme, its real advantage was enjoyed by the farmers, not the power producer. The farmers had an opportunity to solve 100 percent of their waste disposal problem, while the power producer was able to alter its fuel requirements by only (at best!) 10 percent.

(continued from page 111)

than coal to provide the same amount of thermal energy. Under these circumstances, any plant that requires 100 units (by weight) of coal per hour would require 170 units of wood. In a large facility, wood can only supplement the coal supply.

A second challenge in supplementing the coal supply with biomass has to do with the fact that the boiler and associated hardware are expensive and designed specifically for coal. As a fuel, biomass does not have the same characteristics as coal. Many potential bio-mass fuels have higher amounts of potassium and chlorine than coal, for example, and these elements pose potential problems for the coal-fired power-plant operator. Elevated potassium levels are associated with increased ash deposition on the heat exchanger, and when this occurs, the efficiency with which thermal energy is transferred from the boiler to the water is reduced. Reductions in the rate of heat transfer mean higher fuel costs and increased emissions. Increased levels of chlorine in the products of combustion lead to increased corrosion. Grasses, for example, have higher levels of potassium and chlorine than most woody fuels, and so wood is preferred to grass. But coal is, in this sense, better than wood. Retrofitting a boiler to run on large percentages of biofuels is often too expensive, even when it is possible, and so an upper limit of 20 percent biomass is observed by most cofiring operations. In large utility-scale boilers, the limit is usually much lower, typically less than 5 percent; this seems small, but because these boilers operate on such a large scale, 5 percent biomass still means large deliveries of biomass fuel.

The Tennessee Valley Authority, one of the largest power producers in the United States, cofires biomass at Plant A of its Colbert Fossil Fuel Plant, a complex that houses two pulverized coal-fired power plants. The Colbert facility is located in Cherokee, Alabama. Plant A, which began operation in 1955, has a rated power output of 800 MW and burns 320 short tons (290 t) of coal per hour when operating at full power. Plant A was an early experiment in biomass cofiring. It cofires wood. (Plant B, which has a rated power output of 500 MW, does not cofire wood.)

Pulverized-coal technology is an old and widely used method of burning coal on a large scale. The fuel, as the name implies, is pulverized to a very fine consistency prior to burning. Because the coal particles are so small, they heat rapidly upon entering the boiler and burst into flame. Wood does not have to be ground to as fine a consistency as the coal. In order that it burn at the same rate as the

Colbert Fossil Plant showing the coal and biofuel belt with the plant in the background. *(Photo by Joel Love; Tennessee Valley Authority/Colbert Fossil Plant)*

coal, the wood must be ground into particles that are, on average, no more than one-fourth of an inch (0.6 cm) in diameter.

Preparation of the wood is made easy for the Colbert Plant operators, who receive most of the wood in the form of sawdust produced by three furniture manufacturers. There are some larger pieces of wood mixed with the sawdust, and these are removed by passing the wood waste through a device called a trommel screen. Once the larger pieces of wood waste have been removed, the wood fuel is weighed, because no more than 4 percent of the fuel stream (measured by weight) can be wood. The wood is then loaded onto a moving belt, which carries it to the point where the coal is loaded. Next, the belt transports the coal-wood mixture to a bunker for storage. Later, just prior to feeding the fuel to the boiler, the mixture is passed through a pulverizer, a device that reduces the coal to a powderlike consistency.

The contribution of wood to plant output at the Colbert facility is evidently small. Nevertheless, the Colbert plant operators characterize the project as a success. Burning wood helps the operators to reduce sulfur dioxide emissions, which are always a concern at a coal plant, and fuel costs are slightly reduced as well, because the wood waste is free. The project is also of value to the furniture makers, who, if they did not send the wood waste to the power plant, would be burdened with the cost of disposing of it. Finally, because the amount of wood consumed is small compared to the amount of coal consumed, it can be mixed directly with the coal. As a consequence, no expensive modifications of the plant were needed to bring the system into operation. This system is typical of many power-plant systems that cofire biomass.

PART III

Environmental and Policy Considerations

Large-Scale Use
of Biomass

For many, it is more satisfying to look at a forest and imagine a wildlife sanctuary or a source of wood to build homes and furniture—or to look at a vast expanse of grain and see a source of food or a way station for migrating birds—than to look at forest and field and see "biomass," which is just another term for "fuel for the fire." But fire is as important to human life as homes, furniture, and food, and biomass is one method of feeding the fires upon which so much depends. The advantage of seeing plant and animal matter as fuel is that it enables one to apply the same physical principles that are applied to other energy sources. This elemental way of looking at matter has enabled engineers and scientists to make efficient use of each energy source to which the physics and chemistry of combustion has been applied.

In this chapter biomass is examined at very elemental levels. First, biomass is examined as a form of solar energy. In particular,

Los Angeles freeway. Meeting the demand for fuel on this scale solely with biofuels is, at present, impossible. *(En Tien Ou)*

how much solar energy can be converted into biomass? Because the rate at which the Sun supplies energy to a particular area on Earth's surface can be measured, knowing how much solar energy can be converted to biomass reveals the maximum rate of formation of fuel. This leads to the second question to be considered: What is the minimum amount of land needed to produce a given amount of fuel? (Most of the numbers in the first section of this chapter and some of the numbers in the second are taken from an article entitled "Biomass for Energy: Supply Prospects," references to which can be found in the Further Resources appendix of this volume. Other analyses use somewhat different numbers and assumptions, but all the analyses arrive at the same general conclusion: Plant matter fails to convert most of the incident solar energy into chemical energy, a fact with important consequences for agriculture and land use.)

BIOMASS AS A FORM OF SOLAR ENERGY

Biomass is a form of solar energy in the sense that plants convert solar energy into chemical energy, and biomass solves the problem of "intermittency" in the supply of energy from the Sun. At any given location on Earth's surface, the supply of solar energy is unsteady. Consequently, it is not always available when it is needed. No solar energy is available at night, for example, and the rate at which solar energy is supplied is much reduced on cloudy days. The reason that intermittency is a major barrier for those seeking to make direct use of solar energy is that there is at present no way to store large amounts of electricity. One can produce electricity using conventional solar power technologies for as long as the Sun shines, but when the Sun ceases to shine, there is often little to do but wait until the Sun shines again.

Biomass is an exception. Because solar energy is stored as chemical energy, it can be released when the biomass is burned, and biomass can be burned at the convenience of the user. Because biomass solves the problem of intermittency, the next step in determining its value is to estimate how efficiently it stores solar energy—that is, when the Sun shines on a field of plants, how much of its energy is stored in the tissue of the plants as chemical energy? Mathematically, a rough estimate of biomass's efficiency as a storage medium is simple to obtain. Each step in the conversion process can be expressed as a number representing the conversion efficiency of that step. The numbers are then multiplied together to produce an estimate of the efficiency of the overall process. The following "efficiency equations" are estimates of the percentage of the Sun's energy that can, under the best of circumstances, be converted into biomass. First, however, it is necessary to review some facts about the way that green plants convert solar energy into chemical energy.

The mechanism by which plants convert sunlight into chemical energy is called photosynthesis. Sunlight provides the energy needed to power the photosynthetic reaction, which, roughly speaking, consists of converting carbon dioxide and water into glucose and oxygen. A good place to begin, therefore, in any description of this phenomenon is with the nature of sunlight.

Visible light is a form of electromagnetic energy, and as such it shares many properties in common with other types of electromagnetic energy, including X-rays and radio waves. In particular, all types of electromagnetic energy travel in waves. Scientists identify each type of electromagnetic energy by its wavelength, which is the distance from the peak of one wave to the peak of the next. For visible light, the distance from peak to peak is measured in millionths of a meter. Different wavelengths of visible light can be distinguished by the human eye because different wavelengths are perceived as different colors. (Visible light occupies only a small part of the spectrum of all electromagnetic waves, and electromagnetic waves with wavelengths outside the narrow region of the electromagnetic spectrum containing visible light cannot be detected by the eye.)

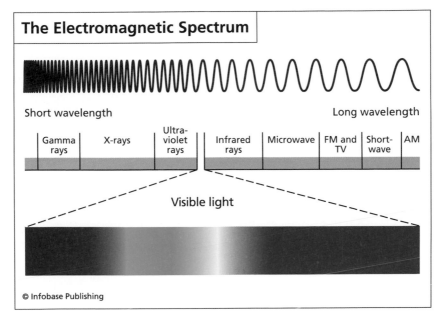

The Electromagnetic Spectrum

Short wavelength   Long wavelength

| Gamma rays | X-rays | Ultra-violet rays | Infrared rays | Microwave | FM and TV | Short-wave | AM |

Visible light

© Infobase Publishing

The Electromagnetic Spectrum: Visible light occupies only a small part of the spectrum.

When all colors (or wavelengths) of visible light are mixed together, the result is ordinary (white) sunlight. The different wavelengths of visible light can also be unmixed. A rainbow, for example, will separate the colors of which sunlight is composed, making each previously invisible color visible to the observer. In a rainbow the shortest wavelengths appear toward the inside of the rainbow. The farther to the outside of the rainbow one looks, the longer the wavelengths one perceives. The bands of color in a rainbow include, in order of increasing wavelength, violet, indigo, blue, green, yellow, orange, and red.

There are other electromagnetic rays present in sunlight that are not visible to the eye. Ultraviolet light, for example, has a wavelength that is shorter than violet, and infrared light has a wavelength that is longer than red. Although these electromagnetic waves have many properties in common with visible light waves, neither infrared nor ultraviolet electromagnetic waves will trigger the photosynthetic reaction.

Photosynthesis is only activated by electromagnetic rays with wavelengths between 0.4 and 0.7 microns. (One millionth of a meter equals one micron.) Red light has a wavelength of 0.7 microns, and violet has a wavelength of 0.4 microns. These facts are important because of the following:

> FACT #1: Only about 50 percent of the solar energy reaching Earth's surface consists of electromagnetic waves with wavelengths between 0.4 and 0.7 microns.

Even when there is no dust on the leaves of the plant and nothing is shading the plant, only some of the light in the 0.4–0.7 micron band will trigger photosynthesis. The rest of the light will be reflected off the leaves, or pass through the leaves, or be absorbed by materials that are not photosynthetic. To be specific:

> FACT #2: About 80 percent of the visible light shining on a plant will trigger a photosynthetic reaction.

Only some of the light energy that participates in the photosynthetic reaction is stored in the glucose molecule. Most of the light energy involved in photosynthesis will dissipate into the surrounding plant matter and contributes nothing to the production of biomass. In particular:

> FACT #3: About 28 percent of the energy carried by the light rays that participated in the photosynthetic reaction is stored in glucose.

Finally, the plant stores chemical energy for its own use—not, of course, for the use of others. Much of the energy stored in the glucose molecule will be used by the plant before it can be released through combustion:

> FACT #4: About 40 percent of the energy stored during photosynthesis is used by the plant to sustain itself. Or to put it another way: About 60 percent of the energy stored in a plant's tissues will remain stored until the biomass is ready to be burned.

The percentages in the preceding four facts can be multiplied together to obtain the maximum percentage of sunlight that is stored in the form of biomass. Beginning with 100 percent of the energy contained in sunlight and multiplying 100 by the numbers obtained in the preceding four facts yields:

$$100 \times 0.60 \times 0.28 \times 0.80 \times 0.50 = 6.7$$

This estimate shows that no more than 6.7 percent of the energy received by a plant from the Sun will be converted into biomass. Sugarcane, for example, can, under ideal conditions, store energy from the Sun at roughly this percentage.

Plants are sometimes divided into two groups when it comes to photosynthesis. Those plants that are more efficient at storing the Sun's energy as chemical energy are called C_4 plants, and the 6.7 percent figure applies to them. Many plants, including most woody plants, are not as efficient as sugarcane at storing solar energy. It is only to these less efficient plants, called C_3 plants, a class that includes most of the plants that grow in temperate climates, that the next two facts apply.

C_3 plants are unable to make use of intense light. They become "light saturated" in the sense that after the light has reached about 70 percent of its maximum intensity, the photosynthetic reaction is already proceeding at its maximum rate. When the intensity of the light exceeds this 70 percent threshold, the "excess" light has no effect on the rate of photosynthesis.

FACT #5: C_3 plants are able to make use of only about 70 percent of the light that shines on them when the light is shining at full intensity.

The value of C_3 plants as an energy-storage medium is further reduced because they undergo a process called photorespiration, which occurs simultaneously with photosynthesis. Photorespiration further reduces the efficiency of photosynthesis because its effect is to undo some of the work of photosynthesis.

FACT #6: Photorespiration in C_3 plants reduces the rate at which photosynthesis stores light energy by an additional 30 percent, or to put it another way: Photorespiration reduces the efficiency of plants in converting solar energy to chemical energy to 70 percent of what it would otherwise be.

To compute the maximum efficiency of C_3 plants, begin with the 6.7 percent efficiency of C_4 plants and multiply that percentage by the numbers in facts 5 and 6:

$$6.7 \times 0.70 \times 0.70 = 3.3$$

So woody plants, as well as many others, will convert no more than about 3.3 percent of the energy in sunlight into chemical energy.

Therefore, under ideal conditions, C_4 plants will convert no more than about 6.7 percent of the incident sunlight into chemical energy, and C_3 plants will convert no more than about 3.3 percent into chemical energy. To put it another way, *under ideal circumstances* C_4 plants will fail to convert about 93.3 percent of the incident light energy into chemical energy, and C_3 plants will fail to convert about 96.7 percent of the incident light energy into chemical energy. The next step is to determine what these percentages mean in terms of land use.

HOW MUCH LAND FOR HOW MUCH FUEL?

Because biomass can be described as an energy-storage medium, and because each plant requires a certain amount of land to grow, statements about energy production from biomass can be interpreted as statements about land use. The percentages obtained in the preceding section can be used to estimate the amount of land required to produce a given amount of energy when the energy is obtained from biomass.

The amount of land required to store a given amount of solar energy depends to some extent on the crop and on the location. Consider the region around Des Moines, Iowa, which lies at the heart of one of the most productive agricultural regions in the United States. (The following calculations are done only with metric units, since these types of measurements are generally expressed only in metric units.) During June, Des Moines receives on average about 149,000 megajoules (MJ) per hectare of energy from the Sun each day. The growing season around Des Moines is roughly 160 days long. An upper estimate for the amount of solar energy that an acre of land receives during the growing season, would, therefore, be 160 days times 149,000 (MJ) per hectare per growing season, or 23.8 million MJ per hectare (160 × 149,000 = 23,840,000). Keep in mind that this is an overestimate, since the amount of energy received from the Sun in June would exceed, on average, the amount of energy received from the Sun toward the beginning and the end of the growing season.

The calculations in the previous section show that C_4 plants would, therefore, convert 6.7 percent of 23.8 million MJ of solar energy available per hectare per growing season into chemical energy for an "energy yield" of about 1.6 million MJ per hectare per growing season:

$$0.067 \times 23,800,000 = 1,594,600$$

A farmer who plants energy crops near Des Moines could not do better than this. The energy content of a field of C_3 plants would

As demand for biofuels increases, difficult decisions about land use will have to be made. *(iStockPhoto.com)*

produce an energy yield of about 787,000 MJ per hectare per growing season, since these plants are only about half as efficient at converting solar energy into chemical energy as are C_4 plants.

Energy yields can be expressed in terms of metric tons of biomass per hectare by using the heating value of a sample of biomass. The heating value of a particular sample of vegetation can be obtained by thoroughly drying the sample and then burning it under carefully controlled laboratory conditions while measuring the amount of thermal energy produced. A rough value for the energy content of a C_4 species is about 17,500 MJ per dry metric ton. (The relatively small difference between higher and lower heating values can be ignored here since the estimates used in this chapter are so rough.) By dividing the heating value into the energy yield per hectare, one obtains an upper estimate for the amount, measured by weight, of C_4 vegetation that can be produced per unit of land. The result is about 91.4 metric tons of C_4 vegetation per hectare:

$$1,6000,000 \div 17,500 = 91.4$$

About half this much, or 45.7 metric tons per hectare, is an upper bound for the maximum amount of C_3 vegetation that can be produced. Both numbers overestimate the amount of biomass currently recoverable using the best available technology by several hundred percent.

As a practical matter, many factors other than sunlight limit the rate at which vegetation grows. The availability of water and soil nutrients and the amount of carbon dioxide in the air are also critical factors, and a plant can only grow as fast as the rate permitted by the least available resource. The limiting resource varies from location to location, and it is not always evident why plants fail to grow at their optimum rates. Many early global warming models, for example, predicted that forests would grow faster as the level of carbon dioxide in the atmosphere increased, thereby moderating the effects of rising carbon dioxide emissions, but this growth spurt has, for the most part, not been observed. The reason is that the model used by scientists was too simple, too unrealistic. In many cases, a shortage of a critical resource other than carbon dioxide limits the growth of a particular forest, in which case, increasing levels of carbon dioxide can have no effect on growth rates.

Other factors, such as insect pests and plant diseases, further limit the production of biomass. And not all vegetation can be recovered. Often, for example, the root system of a plant is a significant portion of a plant's mass, and it would be too disruptive to pull the root system out of the ground even if it were practical to do so, since removing all of the plant matter leaves the field vulnerable to erosion and depletes the soil of its nutrient content. As a practical matter, one-third or even less of the theoretical maximum amount of biomass should be recovered.

⏻ Siting Biorefineries

Because ethanol and biodiesel feedstock are not, when compared to fossil fuels, energy rich, it makes little economic sense to transport large volumes of feedstock for processing. Too much fuel is expended transporting too little energy. As a consequence, ethanol refineries and biodiesel refineries are generally much smaller than petroleum refineries. By way of example, the largest refinery in the United States, the ExxonMobil Refinery in Baytown, Texas, can process 562,600 barrels of crude oil per day, but all 134 ethanol biorefineries currently in operation in the United States as of December 1, 2007, have a combined capacity of about 472,000 barrels of ethanol per day, about 84 percent of the output of ExxonMobil's Baytown refinery. The smaller ethanol capacity reflects the smaller supply of ethanol. The large number of ethanol refineries reflects the fact that it is more economic to build numerous small refineries near the fields where the feedstock is grown than to build a few large refineries and transport the feedstock longer distances. This is, again, a consequence of the lower energy value and more dispersed nature of ethanol feedstocks. The same statements are true of biodiesel feedstocks, and similar statements are true of biomass feedstocks in general. There is nothing to be done about this. It is, at its most basic level, a reflection of the small percentages of solar energy that are captured and stored in all types of biomass.

The large number of small ethanol refineries (as well as biodiesel refineries) is an inefficiency characteristic of the industry—inefficient in the sense that small-scale refineries are more expensive to build and operate because they fail to benefit from economies of scale. Even at this writing, when the price of oil per barrel can fluctuate significantly, ethanol and biodiesel producers remain dependent on subsidies from federal, state, and local

Currently, practical maximums for C_4 plants growing in temperate regions seem to be limited to 2 or 3 percent of the solar energy that shines on a field during the course of year; for C_3 plants, the practical maximum is about 1 percent. As strains of plants are

Train leaving a grain elevator in the Midwest. How much oil should be consumed transporting corn to ethanol refineries? *(Andrew Penner)*

governments to remain in business. But it is reasonable to expect that over the medium term, the average price of oil will not diminish and may even continue to climb in response to additional demand from newly industrialized countries and (perhaps) the continued decline in value of the U.S. dollar in relation to other currencies. Meanwhile, ethanol and biodiesel producers continue to increase the efficiency with which they produce their products. It would not be surprising, therefore, if these biofuel products eventually become competitive with oil, but that day has not yet arrived.

improved and harvesting technologies are improved, it is reasonable to expect that the yields obtained will also continue to improve. Steadily increasing production has been the history of agriculture for generations. There is no reason to suppose that farmers and agriculture firms have reached a point where production has peaked. Nevertheless, the numbers obtained in this section show that if biomass is grown on a scale that is commensurate with the demand for energy, hard decisions about land use lie ahead.

Alternative Reasons for the Consumption of Biofuels

There are many biofuel feedstocks that have not been discussed in this volume so far. These include landfill gas, used cooking oil, municipal solid waste, and construction debris. Many other examples exist. What they have in common is that the total energy value of each such source is relatively small, a fact that may reflect the low heating value of the fuel or the small total supply. In either case, these fuels are economically marginal. Nevertheless, stories about these fuels are regularly featured in the media—and with good reason.

Many economically marginal biofuel resources may still be worth developing, either because they serve important niche markets or because their use contributes to the solution of an otherwise difficult environmental problem. That is the case with the energy sources considered in this chapter. Although small in terms of their potential contribution to the total energy supply, their value is best measured in other ways. Three cases are described. The first section describes the use of a biofuel as a form of pollution control. The second section describes the way that the pulp and paper industry has learned to reduce waste and increase productivity by making better use of the principal byproduct of pulp production. The third section describes why it may be desirable to harvest deadwood for power production even when the economics of the operation may indicate otherwise.

A FORM OF ENVIRONMENTAL PROTECTION

The United States has approximately 1,800 municipal solid-waste landfills in operation, and together they form the final destination for 60 percent of the nation's municipal solid waste. Landfill design criteria have evolved over the years as a number of groups, including landfill operators and government researchers, have collected data on landfill performance. As a result of this research, new government-imposed regulations specify how landfills may be built and operated. The effects of these regulations have been dramatic. Newer landfills are less likely to pollute the groundwater, for example, or to permit methane, an odorless and highly flammable gas, from seeping through the ground and escaping from the landfill site. Landfills are now safer and less polluting than they have ever been.

Just as regulations have changed landfill design, they have also changed the contents of the landfill. Regulations defining "hazardous" waste have become stricter as have the regulations that specify how it should be handled. Definitions governing which materials should be recycled have become broader, and as a consequence of these new definitions and regulations, the composition of the waste stream has changed. Landfills are, therefore, works in progress. Each layer of waste reflects the technology at the time the layer was deposited and perceptions about what was dangerous, recyclable, or valuable. Each landfill is unique.

Energy production using landfill gas can make only a small contribution to the grid, but it could make a large contribution to antipollution efforts at the landfill. *(Rob Hill)*

One characteristic that all landfills have in common is that they are inhabited by bacteria that decompose some of the material in the landfill. Methane is a byproduct of this process of decomposition, and a great deal of methane is produced in this way. In fact, in the United States, landfills are the largest single source of methane produced as a result of human activity. Landfill emissions constitute 25 percent of the nation's total human-generated methane emissions.

Landfill gas is the term used to describe the gas emitted by a landfill, and, when measured by volume, methane constitutes about half of all landfill gas. (Precise percentages vary from one landfill site to the next.) The non-methane component is largely carbon dioxide and water vapor. Usually, there are also small amounts of hydrogen sulfide, a bad-smelling gas that, when it escapes, elicits complaints from a landfill's neighbors. There are also trace amounts of volatile organic compounds, which contribute to the formation

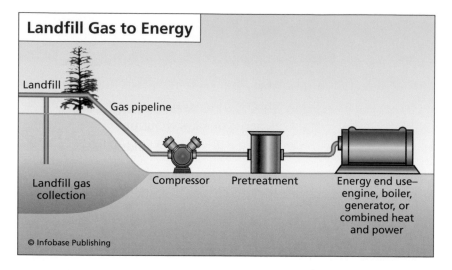

Landfills might be easier to site if the heat and electricity that they produced were sold at a deep discount to the facilities' neighbors. *(Source: EERE)*

of ozone, a constituent of smog, and landfill gas often includes small amounts of hazardous air pollutants such as cancer-causing benzene and toluene.

In addition to being a potential health and safety hazard, landfill gas is a potent greenhouse gas. Methane, its principal component, contributes to global warming in the same sort of way that carbon dioxide does. The difference is that methane's effect on the heat retention properties of the atmosphere is about 20 times as potent as that of carbon dioxide—in other words, the release of one unit of methane is equivalent to the release of about 20 units of carbon dioxide. Because a large landfill can emit thousands of tons of methane each year, the cumulative effects of the nation's landfills on the atmosphere are significant.

Although landfill gas is a potential safety hazard and a source of air pollutants, it is also a potential resource. Because of its methane content, landfill gas burns. It has a heating value that, depending on the characteristics of the site where it is produced, varies from 300 to 600 Btu per cubic foot (11–22 MJ/m^3). By way of comparison, this is between one-third and two-thirds of the heating value of natural gas. To make use of this energy source, about 400 landfills have installed equipment to gather the gas and burn it, thereby generating heat and/ or power. In this way, a large landfill can produce sufficient fuel to power an electricity-generating station with an output of a few megawatts, enough to supply the needs of a few thousand homes. (Precise amounts depend upon the quality of gas at the landfill, the size of the landfill, and other details that are site-specific.)

Such a small generating station may be significant to the landfill operator as a source of income, but its contribution to the grid is not significant. In most situations, if such a landfill were to suddenly stop producing electricity, other power producers would easily make up the difference, and consumers would not even notice. The real value of a landfill power facility lies less in the power that it produces and more in the fact that *as it produces power* it solves most of the problems associated with landfill gas. By burning landfill gas, the foul-smelling hydrogen sulfide molecules are destroyed as are the cancer-causing benzene and toluene and the volatile organic compounds. The methane, which is both a safety hazard and potent greenhouse gas, is converted to water vapor and carbon dioxide. The carbon dioxide is, of course, still a greenhouse gas, but it is much less potent in its effects than methane. The power production scheme, therefore, converts a potential hazard, landfill gas, into a valued resource, electricity.

About three-fourths of the nation's landfills do not convert the chemical energy of landfill gas into electricity or heat for buildings or industrial processes. As of January 1, 2008, about 435 landfills simply flared the gas—that is, they burn it in the open air without attempting to harness the resulting thermal energy. In this way, they eliminate most of the negative impacts associated with landfill gas, but their strategy fails to use the gas to produce a product—either heat or electricity—for which there is demand. The fact that more landfills prefer to simply flare landfill gas than convert it into a useful product is a good indication that the economic benefits of heat and power generation are modest.

The practice of converting landfill gas to heat and power illustrates one of the peculiar characteristics of many schemes to harness low-energy biofuels (low in the sense that either they contain little energy on a per unit mass basis or their supply is too limited to become an important source of energy). The principal advantage of using landfill gas to generate heat or power is that it solves almost 100 percent of an air pollution problem. To represent landfill gas-to-electricity schemes as power generation technologies is, in a sense, to miss the point. Their real value is as pollution-control measures. The heat and power may be sufficient to pay the costs of the pollution-control measures—ideally, they may even generate a profit—but if a power generation scheme were to prove unprofitable, it might still make good environmental sense to go ahead with such a project.

This view has not been widely accepted. Today, almost half of municipal landfills continue to vent landfill gas directly into the atmosphere.

THE PULP AND PAPER INDUSTRY

The U.S. pulp and paper industry is the nation's leading industry in the use of biofuels. For decades it has demonstrated a keen interest in developing new technologies that would permit it to use more biomass for energy production.

In the United States, pulp and paper manufacturing is big business and an important source of export dollars. U.S. companies supply more than one-fourth of the world's paper needs. This level of production requires a large number of workers. In 2006, almost 128,000 individuals were employed in the mills that produce pulp, paper, and related products. The pulp and paper industry is also the third-largest user of energy among all manufacturing sectors, surpassed only by the chemical and metallurgical industries. Roughly 60 percent of the pulp and paper industry's energy needs are met with various types of biofuels.

Paper mill. The pulp and paper industry has long led the way in utilization of biomass energy. *(iStockPhoto.com)*

The papermaking process usually begins by applying onto a screen a liquid-solid mixture consisting chiefly of water and cellulose fibers. (Recall that cellulose is the chief constituent of cell walls.) The water drains through the screen, but the fibers and any additives are trapped. Additional steps are taken to remove even more water. Next, the fibrous layer is forced through a series of presses and rollers, and as they pass through this complex processing equipment, the fibers form a mesh. What emerges is paper. Of most interest for the purposes of this volume, however, is what happens before the liquid-solid mixture is applied to the screen—specifically, the mechanism by which the fibers, called pulp, are produced.

The main ingredient of pulp is wood. Paper can also be recycled to produce pulp for new paper, but recycling paper degrades the pulp contained in the paper, and there is a practical limit, currently around 50 percent, for the amount of recycled material that can be incorporated into new paper. For reasons of economy or quality, many manufacturers currently operate at less than the 50 percent limit, which is why the principal constituent of paper is still wood.

To make wood into pulp suitable for paper manufacturing, it is first necessary to strip off the bark. (The bark and other residues generated during the process may be, and often are, burned to help meet the energy needs of the factory and to dispose of a huge quantity of waste.) The wood is then chipped into small, thin pieces. At its most basic level each wood chip consists of fibers bound together by a gluelike substance called lignin and some other materials. This is the nature of all wood. In large measure, pulp manufacturing consists of destroying the cohesive effect of the lignin and separating the fibers. There are several ways to accomplish this, but by far the most common method is called the kraft process. It works as follows:

A soaking solution, called white liquor, is mixed with the wood chips in a large vessel. The soaking solution consists of water and the chemicals that are used to dissolve the lignin. Once the lignin has been dissolved, the soaking stage of the process is complete. At this point, the plant operator has two products, one consisting of pulp and the other consisting of the soaking solution together with the dissolved lignin. Pulp constitutes about 50 percent of wood, and the kraft process is successful in separating most of it. Most of the lignin and other plant material are in the soaking solution, which, because it has changed color, is now called weak black liquor. (The word *weak* refers to the fact that the principal component of the liquid is water.) The pulp is separated and undergoes additional processing in order to prepare it for the papermaking process already described. But the black liquor, which contains a great deal of burnable material as well as the chemicals from the white liquor, is also processed in order to recover the chemicals so that they can be used again. It is the black liquor that constitutes the biofuel of interest here.

The weak black liquor is made into a thicker solution by evaporating much of the water that it contains. The result is a solution that contains anywhere from 60 to 80 percent solids. It is called strong black liquor, and it is burned in a boiler. The combustion of black liquor supplies much of the energy used to power the pulp-manufacturing process. (Often the output from the black liquor is supplemented with wood residues and fossil fuels, but industry-wide, the majority of the thermal energy used to generate the steam and electricity that the pulp mill needs comes from wood residues and black liquor.)

The chemicals used in the soaking process do not burn; they melt. As the black liquor is burned, these chemicals drip to the floor of the boiler and flow out through holes in the bottom. They are recovered, further processed, and then used to create a new batch of white liquor so that the process can begin again. If the black liquor was not processed, a pulp mill would have to continually purchase new chemicals to manufacture the white liquor; it would also be faced with the problem of disposing of the black liquor in an environmentally responsible way, and it would have to acquire another fuel source to generate the heat produced by burning the black liquor. Processing the black liquor solves all three problems simultaneously.

Prior to the oil embargo of 1973, energy of all kinds was relatively inexpensive—not just inexpensive by today's standards, but inexpensive by the standards of the time—and so pulp mills had little interest in optimizing their use of energy and, in particular, their use of biofuels. In 1972, waste wood and black liquor constituted about 40 percent of the energy used at a typical pulp mill. As energy prices have risen, however, pulp mill operators have worked hard to increase efficiencies and to make better use of biofuels. Now, the contribution of biofuels to the energy needs of the pulp industry has increased to about 60 percent of the total. Today, new

Gasification Technology in the Pulp and Paper Industry

The primary reasons that black liquor is burned in pulp mills are to recover the chemicals suspended within it and eliminate a source of pollution. The thermal energy produced by burning the black liquor is a valuable by-product of this recovery process. Nevertheless, black liquor is one of the largest sources of biomass energy in the United States. The technical difficulties involved in burning black liquor so as to maximize the three goals of chemical recovery, pollution control, and thermal energy production have not been completely solved. One problem is that the thermal efficiency of the process is not very high. One way to increase thermal efficiency while decreasing emissions is to gasify the liquor.

The idea behind gasification is to convert the black liquor to a cleaner-burning flammable gas. The gasification process begins with the organic compounds in black liquor. These consist of the lignin and other nonpulp constituents that were separated from the pulp. Chemically, these materials are mostly carbon (chemical symbol C). The carbon is heated in the absence of air. Without oxygen, the carbon will not burn. Instead, the hot carbon is exposed to steam (chemical symbol H_2O), and the result is hydrogen gas (chemical symbol H_2), which is highly flammable and has a high heating value, and carbon monoxide, which is represented by the chemical symbol CO. In symbols, the reaction is expressed as follows:

$$C + H_2O + heat \longrightarrow H_2 + CO$$

(The arrow shows the direction of the reaction. It begins with carbon, water, and heat and ends with hydrogen and carbon monoxide.)

and innovative technologies that increase energy production while simultaneously decreasing emissions are being deployed throughout the industry. The successes of the pulp industry have not attracted much attention because the resources and requirements of

The carbon monoxide can be burned, but it is more efficient to let it, too, react with steam to form hydrogen gas and carbon dioxide (chemical symbol CO_2). The resulting reaction is expressed as follows:

$$CO + H_2O + heat \longrightarrow H_2 + CO_2$$

The extra work involved in converting black liquor to hydrogen gas can be justified in several ways. First, by burning hydrogen gas, emissions are reduced to almost zero, since hydrogen combines with oxygen to produce water as its sole product. (Practically, the process is not as simple as the preceding description implies. Some pollutants, for example, still manage to find their way into the combustion chamber. Despite this qualification, emissions of the most common pollutants can be reduced by about a factor of 10—the exact amount depends on the details of the particular gasification process.) Another advantage of gasification is that hydrogen gas can be burned with a higher thermal efficiency than can black liquor. Increased thermal efficiency means that the operators are able to produce the same amount of power with less fuel. Finally, the process by which the chemicals in the soaking solution are recovered for further use is enhanced in the gasifier, leading to a safer and more economical recycling method.

These gasifiers are only now being introduced in the industry. Gasification technology offers the possibility of converting pulp and a wide variety of other biomass fuels, each with its own characteristic chemistry, into a single clean-burning fuel. The pulp industry is leading the way.

the industry are unique to the production of pulp. Nevertheless, the industry remains one of the biggest success stories in the responsible use of biofuels.

FIRE SUPPRESSION

Decades ago in the United States, the idea that fires had no place in forests began to gain broad acceptance. The complete suppression of forest fires became public policy and was practiced on all lands, public and private. Many young people spent a summer or two during their college years working for the federal government as fire spotters, whose job it was to identify new fires as quickly as possible so that firefighters could be sent to the scene to contain and ultimately extinguish the fire. Fires were perceived as alien to the forest environment.

As indicated in chapter 1, for thousands of years aboriginal peoples in many areas of the world routinely set the forests in which they lived on fire. Setting forests ablaze accomplished several goals at once: The ashes of dead plants fertilized the lands and opened spaces for new vegetation to grow; the new vegetation provided food for the animals upon which the aboriginal peoples depended; fires controlled insect pests that damaged the forests as well as others that were potentially injurious to humans; and burned forests were easier to traverse. Had they not set the forests afire, they would have discovered that given enough time the accumulating dead vegetation would have provided fuel for fires that were much larger, more destructive, and more dangerous than those that they routinely set. Even in the absence of communication, many aboriginal peoples arrived at the same conclusion, and so many of the world's forests smoldered almost continuously for millennia. So if fire was not "natural" to the forest, it was, until recently and in many areas, very familiar. One proof of this statement is that some plants—the jack pine of North American forests being the most frequently cited example—require fire in order to reproduce. (The jack pine's cones

Fuel treatments can make only a small contribution to the energy supply but can make a large difference in efforts to eliminate catastrophic forest fires. *(National Park Service)*

must open before it can reproduce, and the cones will not open until exposed to the heat of a fire.) Fire suppression led to habitat loss for the jack pine and for some of those creatures that depend upon the tree.

Today, many forests, both public and private, can be fairly characterized as having been damaged by the well-meaning policy of fire suppression under all conditions. Trees are increasingly crowded, and the intervening spaces increasingly filled by dry, dead vegetation, making the prospect of highly destructive and dangerous fires ever more probable. These points are now widely acknowledged, and discussion has now focused on what to do about the hazards created by the old policy. Simply setting the forests alight is not a viable solution, because forests are now too full of fuel, a situation that makes the outcome of any intentionally set fire difficult to predict, and because homes and roads are now located throughout many forests. A more nuanced alternative to either suppressing fire indefinitely or allowing a forest fire to run its course is required.

One sustainable solution involves removing excess biomass from forests and burning it to generate electricity. The value of the approach is that it converts a hazard (poorly managed forestland) into a societal good (electricity). The problem is that, as mentioned elsewhere, the heating value of wood is not especially high, and the costs of removing the wood without undue damage to the forest and then transporting the wood to a power station further reduce the economic attractiveness of this alternative.

Fuel transportation costs can be reduced by building small or medium-sized biofuel-powered generation facilities near places where adequate supplies of fuel wood can be produced indefinitely, but in most cases, this has not been done. There are three reasons. First, small power plants fail to benefit from economies of scale, which is another way of saying that as a general rule it costs more to produce a kilowatt-hour of electricity with a small generating station than it does with a large one, perhaps making a small power plant economically uncompetitive. Second, fuel costs are higher when only individual logs or isolated stands of trees are extracted in a way that does not damage the surrounding forest, and this would be the situation if the wood was removed to reduce the danger of fire. Finally, before a generation station's energy can be used, it must be sent across a high-voltage transmission line, and these have become notoriously hard to build. Property owners and self-described environmentalists often oppose the construction of new transmission facilities, and in any case, private industry has evinced little enthusiasm for building the lines, since they are expensive to construct and their operation and profits are tightly regulated. Connecting a new geographically isolated biofuels plant to the grid would be difficult.

But because converting deadwood into electricity can reduce the extent, and so the cost, of forest fires, a more inclusive view of

Fuel treatments involve reducing the fuel density within the forest, thereby decreasing the risk of catastrophic forest fires. Ideally, the biomass that is removed is converted into a societal good, a source of energy. Conversion to electricity is one possibility; conversion to ethanol is another. *(Source: Government Accountability Office)*

forest management may yet lead those responsible for the nation's forest and the nation's electricity infrastructure to look at power generation as a way of helping to defray the costs of managing large forests. Even when a biofuel-powered plant may not make economic sense as a stand-alone endeavor, it can still make sense as a tool of better forest management. How much biomass can a *fuel treatment* program produce? (Fuel treatment is a term used to denote forest fire hazard reduction.) The authors of "Biomass as Feedstock for a Bioenergy and Bioproducts Industry," a paper already described in earlier chapters and listed in the Further Resources appendix of this volume, estimate that roughly 60 million dry tons of biomass are available for harvest for purposes of fuel treatment. This 60-million-ton figure applies after environmentally sensitive areas were excluded from consideration and when forest areas not currently accessible by roads were excluded. There is, therefore, some potential for electricity production.

When considered in a larger context, the consumption of biofuels can sometimes produce economic, environmental, and social benefits that are unrelated to the electricity that consuming them produces. Before asserting that a particular biofuel is uneconomic, it is always important to evaluate all of the potential benefits associated with its consumption. The consumption of biofuels is in many respects a more complex issue than the consumption of fossil fuels, and evaluations of biofuels need to be performed with that in mind.

Policy, Research, and Biofuels

Fossil fuels have proven extremely hard to displace. Despite efforts to encourage the intensive use of biofuels—efforts that began more than 30 years ago in the United States—the contribution of biofuels to the power generation and transport sectors remains small. Many supporters of biofuels attribute the persistence of the world's fossil-fuel economy to the political power of so-called Big Oil and Big Coal, but the situation is more complicated. To be sure, big energy companies exercise political power commensurate with their wealth and their contribution to the economies of the nations in which they operate, and they have sometimes acted to suppress innovation and change in the energy markets. But it is also true that fossil fuels have certain intrinsic advantages that are very difficult to overcome. Failure to appreciate the advantages of fossil-fuel use leaves one with little understanding of why biofuels have so far failed to capture a larger share of the energy markets.

This chapter examines some of the reasons that fossil fuels continue to dominate the energy markets. It goes on to describe some of the ways that the U.S. government has attempted to create a market for ethanol. Ethanol is emphasized, because in the United States the ethanol market is the most developed of all biofuel markets, and a great deal of money and expertise have gone into creating it. Finally, biofuel technology seems poised to undergo a period of rapid technological change. New developments may make it possible to convert many widely varying biomass feedstocks into a few clean-burning fuels. If this occurs, the biofuel industry may finally make the big difference in the energy markets that its supporters have long promised.

OVERCOMING INERTIA

To understand why the contribution of biofuels to the energy sector remains small, it helps to examine why the contribution of fossil fuels remains large. As is widely acknowledged, fossil-fuel consumption has had far-reaching deleterious environmental effects, and the production and consumption of fossil fuels, especially coal, have resulted in a great deal of human suffering. During the 20th century, for example, 100,000 coal miners lost their lives in coal-mining accidents within the borders of the United States alone. The total number of U.S. miners who lost their lives during this period as a result of their work in the mines is actually much higher when the deaths due to occupational diseases such as black lung disease are taken into account. Coal-mining fatalities of all types continue to occur. But the consumption of coal (and oil and natural gas) continues to grow because—and this is not as widely acknowledged—fossil fuels work so well in so many ways. They are plentiful, have high heating values, are relatively safe to transport, and they are reasonably easy to use.

To understand just how plentiful fossil fuels are, consider the following facts. There is enough coal in the United States to maintain present levels of consumption for centuries. The U.S. Energy

Coal-fired plant in Colstrip, Montana. Today's trillion-dollar energy infrastructure was optimized to make use of fossil fuels. It does not easily adapt to other fuel types. *(D. Hanson and Northern Plains Resource Council)*

Information Administration estimates that world oil production will not peak until around 2037, even though current production is roughly 70 million barrels per day and growing. If a way is found to exploit methane hydrates, vast deposits of methane that are, for the most part, located beneath the sea, there will be enough natural gas to last many centuries, even at much higher rates of consumption than exist presently. No such claims can be made for the supply of biofuels. Because biofuels can be grown, they can be produced indefinitely, but they cannot be produced at rates approaching those of any of the major fossil fuels.

As mentioned previously, the heating values of fossil fuels tend to be higher than the biofuels that they most closely resemble. Gasoline has a higher heating value than ethanol; coal has a higher heating value than wood, switchgrass, or other sources of biomass; and even petrodiesel has a slightly higher heating value than biodiesel.

And fossil fuels are relatively safe to transport, although this characteristic is shared by the biofuels that they most closely resemble.

There is another advantage enjoyed by fossil fuels that is no less important than those already described. It has nothing to do with the physical or chemical properties of these fuels. It is, instead, an artifact of history. Energy companies have built enormous and enormously expensive infrastructures to store, distribute, and consume fossil fuels. These facilities are optimized for fossil fuels, and using these facilities to store, distribute, or consume fuels other than the ones for which they were designed is often difficult or impossible. There is no similar infrastructure to support the use of biofuels. The advantages enjoyed by fossil fuels that are historical in origin are often described by saying that fossil fuels benefit from the inertia created by the infrastructure built to produce, transport, and consume them.

Consider the problems faced by the Southern Company, an important power producer located in the southeastern United States, when they decided to burn switchgrass to produce electricity. Roughly 70 percent of the Southern Company's generation capacity is composed of pulverized coal–fired power plants. (As the term implies, pulverized coal-fired plants use coal that is ground to a very fine consistency. This technology is highly reliable and most large coal-fired power plants depend upon it.) On the face of it, the most efficient way to burn switchgrass in a pulverized coal-fired plant would be to cofire it, and the easiest way to cofire it would be to shred the switchgrass, mix it with the coal, and burn the coal-switchgrass mixture together. Southern Company found this approach impractical.

What Southern Company engineers discovered is that even small amounts of switchgrass—only 5 percent by weight—were sufficient to cause the pulverized coal–switchgrass mixture to form a sticky mass that would not flow. As a consequence, the mixture could not be fed to the boiler using the infrastructure that was already in place to handle the pulverized coal. Barred from using what seemed to be the simplest, most effective method of cofiring the switchgrass, the company tried forming the grass into cubes and feeding the boiler small quantities of switchgrass cubes together with the coal. This also failed. Some of the cubes would inevitably be crushed, the grass released, and the coal-switchgrass mixture would fail to flow into the boiler for the same reason that the first method had failed. The problems encountered by the Southern Company in their experiments are not the result of engineering deficiencies. Just the opposite is true. Southern's system was optimized—designed to run as efficiently as possible—for pulverized coal, and changes in the fuel's composition reduced the system's efficiency. In this case, the efficiency of the fuel-delivery system was reduced to unacceptable levels.

Or consider this example of inertia: A large coal-fired plant will generate more than 100 tons of ash per hour. This ash is often sold as an additive for concrete, but for safety reasons only "pure" ash from coal combustion is acceptable. The reason is that only pure ash has been thoroughly tested for suitability as an additive. Because concrete is often an integral part of some of the largest, most important, and most expensive civil engineering projects, changing its formula is not a matter to be taken lightly. Testing is time-consuming and expensive. These facts also reduce the attractiveness of switchgrass, a source of biomass that is often touted as a fuel of the future, as a fuel for cofiring with coal. The reason is that switchgrass contains significant amounts of ash, and the chemical properties of ash created by burning switchgrass differ significantly from those of coal ash. Any company that decides to cofire switchgrass with coal faces the possibility of a significant ash-disposal problem and a potential loss of revenue from sales of ash to concrete companies.

Finally, consider the difficulty faced by a home-heating oil company that wants to switch to biodiesel. From the point of view of combustion, biodiesel is essentially interchangeable with home heating oil. On the face of it, therefore, there seems no reason to delay. But biodiesel degrades seals in old systems and tends to free sediments that previously lay dormant within the system, causing them to begin to circulate and potentially clog the fuel lines. On

(continues on page 158)

Switchgrass: A Future Fuel

The biofuels sector seems poised for rapid expansion as new biofuel production technologies approach commercialization. Many of the more promising technologies have been discussed in this volume, but it is far from clear which technologies will be in wide use in 10 years. Some technologies will probably gain widespread acceptance, and some will, at least temporarily, be cast aside; the difference between "winners" and "losers" to be determined by a complex interplay among technical, economic, and political factors. Nowhere is this more apparent than in the scramble to find an economic way of using a crop called switchgrass.

Switchgrass was one of the grasses that covered the Great Plains region of the United States prior to the 19th century. It grows rapidly, is very hardy, requires relatively little care, and can grow to a height of 10 feet in a single summer. In many ways, switchgrass is an ideal energy crop, but what is the best way to use it? Currently, the demand for switchgrass is low, because technologies that exploit its value as a fuel source are still in the development stage. The following is a list of three of the more talked-about possibilities:

1. Switchgrass as a feedstock for ethanol. As described in chapter 4, cellulosic ethanol is a technology for converting plants such as switchgrass into ethanol. This conversion can already be done in the laboratory, but cellulosic-ethanol technology has not been adopted commercially because the ethanol produced by this technology would be too expensive to sell. The situation will change, however, if (a) the technology is improved so that cellulosic ethanol becomes economical to produce, or (b) cellulosic-ethanol producers receive more generous government subsidies, or (c) the price of gasoline climbs high enough to make cellulosic ethanol competitive. Of course, these three factors can change independently of one another, and the result is difficult to predict.
2. Switchgrass as a fuel for electricity production. While there have been scattered attempts to generate power with switchgrass, it

tends to be more difficult to burn than conventional fossil fuels. The key to making efficient use of switchgrass seems to lie in improved gasification technology. This technology involves the conversion of (solid) switchgrass into a gaseous fuel that would contain significant amounts of carbon monoxide and hydrogen or perhaps pure hydrogen. The conversion eliminates the problems associated with burning the solid fuel, and also enables plant operators to use a more efficient *combined cycle* technology in which (a) the gas is burned to drive a gas turbine, and (b) the residual heat is used to convert water to steam to drive a steam turbine. Combined cycle technology converts a high percentage of the thermal energy into electrical energy. Gasification technology is expensive, and so far power producers have not rushed to embrace it. But with more research and development, costs will come down, and switchgrass may become more attractive as a fuel for power producers.

3. Switchgrass as a fuel for methane production. The price of natural gas has increased severalfold since the 1990s, but for many applications there is still no better alternative. (Most of the time, natural gas, as it is delivered to the consumer, is almost pure methane.) Biomass-to-methane technology is in its early stages of development, but plans to build a demonstration biomass-to–natural gas conversion plant in Brayton, Massachusetts, were announced in October 2007. The plant will also be able to convert coal and petroleum coke, a high-carbon by-product of petroleum refining, into methane.

Although none of these technologies are in full commercial development yet, there is no fundamental reason why they will not all eventually be scaled up for production. Of course, if they all become commercial, the demand for switchgrass and similar feedstocks may exceed supply, inhibiting the further deployment of the very technologies designed to make use of it.

(continued from page 155)

the one hand, the answer is simple: Replace the seals and clean the system. But many homeowners may be unwilling or unable to make the necessary changes. Most heating oil businesses are reluctant to drop customers or to assume the financial responsibility for modernizing their customers' old heating systems. The result is that the home heating oil companies are blocked from using a large percentage of biodiesel in their product because of its effects on older systems.

Despite claims to the contrary, fossil-fuel energy systems are often efficient in the following sense: They have been manufactured so as to make optimal use of the fuels that they were designed to burn. These design criteria limit the ability of each system's operator to switch fuels after the system has been built. Of course, it is often possible, in theory, to reconfigure a particular system to burn another fuel type, but the change is expensive and the benefits uncertain. In the examples just considered, fuel innovations were blocked at each step—not from want of trying on the part of those involved, but because fuel-switching was often technically difficult or too expensive. Many other examples of inertia exist.

The underlying problem that impedes fuel-switching is that the energy infrastructure is already in place. Money has already been spent. Personnel have been educated in the maintenance and operation of each system. Each of these factors operates to slow the introduction of biofuels as well as other alternatives, even when there are advantages associated with the introduction of these new fuels.

LEGISLATING A MARKET FOR ETHANOL

The U.S. government has worked long and hard to create an ethanol market. The history of this effort is instructive in that it shows both the value and limitations of government support.

Substantial ethanol markets developed without government support during the first half of the 20th century. Ethanol was used

Government policies to create biofuel markets have had numerous unintended effects. *(Architect of the Capitol, U.S. Government)*

as an antiknock ingredient in gasoline during the 1920s, and during the 1930s, a few thousand filling stations located in the Midwest sold gasoline-ethanol blends. This market collapsed shortly after World War II, and virtually no ethanol was produced for use as a transportation fuel from the late 1950s until the 1970s.

The oil crises of the 1970s caused the federal government to reexamine the value of ethanol as a transportation fuel. The nation had huge surpluses of corn, and petroleum supplies were no longer so certain. What followed were a quick flurry of tax incentives, research initiatives, and government studies aimed at encouraging the use of ethanol in the transportation sector.

The first major piece of tax legislation to encourage the use of ethanol production was the Energy Tax Act of 1978. It was supposed to encourage the production of a fuel called gasohol, which was defined as a gasoline-ethanol blend consisting of at least 10 percent ethanol. The government offered a 40-cent per gallon (approximately 11-cent per liter) subsidy for each gallon of ethanol produced. By the standards of the time, this was an extremely generous subsidy. Potential ethanol producers received further encouragement in 1980 when Congress passed the Energy Security Act, which contained loan guarantees for small plants (defined as less than 1 million gallons [3.9 million liters] per year) that would cover up to 90 percent of the cost of the plant. The Energy Security Act further directed the U.S. Agriculture Department and the Department of Energy to prepare a plan to boost ethanol consumption with a goal of replacing 10 percent of the nation's gasoline supply by 1990. (Studies by these two agencies later showed that ethanol production would have to increase to about 11 billion gallons [42 billion liters] per year and would result in a 75 to 85 percent increase in commodity prices. Three decades later, the 10 percent goal has yet to be met.)

An ethanol plant building boom ensued. By the beginning of 1984, 163 ethanol plants were in operation. The Surface Transportation Assistance Act of 1984 increased the ethanol subsidy to 50 cents per gallon (approximately 13 cents per liter), and the subsidy was soon increased to 60 cents per gallon (approximately 16 cents per liter) by the Tax Reform Act of 1984, all to no avail. By the end of 1984, 55 percent of the ethanol plants in operation at the beginning of the year were out of business. The subsidies had not been enough. The falling price of oil on the international markets had made ethanol unprofitable. Even the generous subsidies of the 1980s could not overcome the impact of cheap petroleum. The more fundamental problem, however, was Congress's unrealistic expectation that ethanol could be produced in sufficient quantities to replace significant volumes of gasoline. Given the technology of the time, this was not possible, and no set of laws could change that fact.

The real market for ethanol in the United States was, at least for the next few decades, not as a fuel but as a fuel additive. In 1973, the Environmental Protection Agency issued regulations with the goal of eventually eliminating lead from gasoline. At the time, manufacturers added lead to gasoline in order to prevent the gasoline from igniting prematurely inside the cylinder, a phenomenon known as engine knock. (Engine knock decreases fuel economy and increases engine wear.) As indicated in chapter 4, small amounts of ethanol or a chemical called MTBE can serve to substitute for lead in gasoline. In response to the Clean Air Act Amendments of 1990, the Environmental Protection Agency began to issue regulations to reduce harmful emissions associated with burning gasoline. Emissions could be reduced by adding either additional ethanol or additional MTBE to gasoline. Both strategies were used, but MTBE was favored because it could be mixed at the refinery and shipped through the enormous pipeline infrastructure that links large refineries with many large metropolitan areas; ethanol cannot be mixed with gasoline and sent through the pipeline infrastructure. Instead, it is transported by railcar and mixed just prior to distribution, a strategy that made ethanol uncompetitive with MTBE throughout much of the nation.

An important turning point for ethanol producers came in 1999, when California began the process of phasing out the use of MTBE. Other states soon followed. Ethanol production soared. It was, after all, still benefiting from generous government subsidies that fluctuated in the 50 cent per gallon (13 cents per liter) range, and it faced little competition. This is the situation as it exists today. The real value of ethanol is still as a fuel additive.

Congress, however, has never abandoned the idea that ethanol could become an important transportation fuel. In the 1990s, this belief was incorporated into a piece of legislation called the Energy Policy Act of 1992 (EPAct). EPAct is a complicated piece of legislation that affects many parts of the nation's energy infrastructure, but one aspect of the law sought to reduce petroleum imports by requiring the purchase of alternative-fuel vehicles by operators that maintained large centrally fueled fleets of vehicles in metropolitan areas. The term "alternative fuels" included E85, the ethanol-gasoline blend consisting of 85 percent ethanol; B100, which is the term for pure biodiesel; electricity for battery-powered cars; hydrogen; liquid fuels derived from coal and natural gas; and a few other fuels, none of which have gained widespread acceptance.

The goal of reducing petroleum imports was further bolstered by presidential executive orders. Notable among these was the executive order of December 13, 1996, that focused on the acquisition of alternative-fuel vehicles by federal agencies. As a result, manufacturers began to sell the federal government large numbers of flex-fuel vehicles, the term for cars and light trucks that can burn more than one fuel. In practice, the term often means cars and light trucks that can burn gasoline or any ethanol-gasoline blend. Many flex-fuel vehicles also found their way into private hands. Practically speaking, most of these vehicles burned only gasoline. Indeed, many vehicle operators did not even know that their vehicles could burn alternative fuels, and in any case, even if many of these drivers had wanted to buy a rich ethanol blend, they would not have been able to find it. In 2002, six years after the order was issued, and 10 years after EPAct of 1992, less than 200 filling stations in the United States carried E85.

Congress has continued to pass energy legislation. A more recent effort is the Volumetric Ethanol Excise Tax Credit, part of the American Jobs Creation Act of 2004, which provides a host of subsidies in order to increase the nation's ethanol production, including a 51-cent per pure gallon of ethanol (about 13-cent per liter) tax credit available from 2005 until 2010. Another piece of federal legislation, the Environmental Policy Act of 2005, sought to further bolster the ethanol market by mandating a minimum amount of renewable fuel, which in this case can only mean ethanol, in conventional gasoline—4 billion gallons (15 billion liters) in 2006, a goal that was easily met, and 7.5 billion gallons (28 billion liters) by 2012, a goal that seems within reach. It also attempts to create a market for cellulosic ethanol by seeking to create a minimum 250-million-gallon (950-million-liter) market by 2013, a requirement that depends on technology that has not yet been deployed.

Another executive order, this one dated January 24, 2007, repealed the 1996 executive order and instead began to mandate that federal fleets decrease petroleum usage by 2 percent per year relative to their petroleum usage during fiscal year 2005. The Department of Energy's Office of Energy Efficiency and Renewable Energy acknowledged that meeting this goal would probably require federal fleet operators to decrease fleet size and vehicle miles traveled, which is one more admission that a straight gasoline-to-ethanol switch remains as much out of reach today as it was decades ago. As a fraction of the transportation fuels market, ethanol will remain in the single digits for years to come.

The most recent attempt to create a large market for ethanol is the Energy Independence and Security Act. Passed by Congress in December 2007, this piece of legislation attempts to mandate an increase in ethanol production from 6 billion gallons (23 billion liters) per year in 2006 to 36 billion gallons (136 billion liters) per year in 2022. Perhaps this time the production goals will be met, but as with previous attempts to mandate high levels of ethanol use, this one also depends on the deployment of production technologies that have yet to be commercialized.

For more than 30 years Congress and the executive branch have tried to create a market for ethanol that would make it an important transportation fuel. In retrospect, these efforts serve as a reminder that no matter how generous the tax credits or aggressive the purchasing requirements, one cannot legislate physics.

SCIENCE, COMMERCE, AND BIOFUELS

If it were possible to simply legislate a change in the biofuels markets, that change would have occurred 30 years ago with the Energy Tax Act of 1978. But if elected officials have not yet learned from past experience, the same cannot be said for the scientists and engineers working in government research laboratories and those managing federal research monies.

There are huge gaps between research into fundamental questions about the characteristics of biofuels and the creation of an energy infrastructure capable of manufacturing the fuels in bulk, distributing them, and using them. These are the gaps between theory and practice, and they are keenly felt by the many institutions, public and private, seeking to bring these fuels into large-scale commercial production. Examples of differences between theory and practice have been mentioned throughout this text and include (1) cellulosic ethanol, which, while possible in the laboratory, has yet to become a commercial reality, (2) cofiring switchgrass with coal, which has proven to be more difficult than early supporters had anticipated, and (3) gasification technology, which, while possible in the laboratory, has proven to be difficult to commercialize for a variety of practical reasons. Ideas of how to bridge the gaps that exist between foundational research and large-scale commercial projects have changed over the last half-century. In response, new models for government-sponsored scientific and engineering research have been created.

Changes in the method by which research is conducted began in earnest in 1958 when the U.S. Defense Department established the Advanced Research Projects Agency, later named the Defense Advanced Research Projects Agency (DARPA). DARPA was created in response to the 1957 launch of *Sputnik*, the world's first artificial satellite, by the former Soviet Union. DARPA's task was to identify and investigate new scientific ideas that could be rapidly brought to fruition. Most research efforts were (and are) carried out with high-quality staff working on three- to five-year contracts. There is a strong emphasis on accomplishing the goal within that time frame. Staff members are rotated regularly in and out of DARPA from agencies such as NASA to prevent the onset of

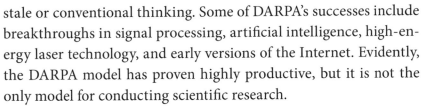

stale or conventional thinking. Some of DARPA's successes include breakthroughs in signal processing, artificial intelligence, high-energy laser technology, and early versions of the Internet. Evidently, the DARPA model has proven highly productive, but it is not the only model for conducting scientific research.

Established in 1950, the National Science Foundation (NSF) was the nonmilitary organization tasked with encouraging research outside the Defense Department. In contrast to DARPA, the NSF emphasized nondirected "pure" research, and through its funding decisions took an approach that was very much the opposite of DARPA's, preferring to support open-ended basic research. But DARPA was so successful that in 1971, at the insistence of President Nixon, a DARPA-like institution was established within the NSF called Research Applied to National Needs (RANN). The idea was that RANN analysts would search for ideas with large economic potential or the potential to transform society in some other fundamental way. Once an idea or technology was identified, an interdisciplinary staff would be rapidly assembled to solve the critical part of the problem—that is, it was the job of the team to bridge the gap between concept and implementation. If successful, the team's solution would be transferred to the organizations best able to make use of it. In any case, successful or not, the effort would soon be brought to a conclusion, and RANN analysts would continue to search for new ideas and applications. Among the projects undertaken by RANN were research into automotive propulsion, fuels, energy conversion, and energy storage.

Although it only used 10 percent of the NSF's budget, the RANN program was unpopular with the NSF directors, who wanted to apply even more of the agency's considerable resources to supporting foundational research. They had little interest in funding the sorts of high risk–high payoff research projects that RANN had been charged with pursuing. They preferred to continue to emphasize basic research as opposed to a more comprehensive results-oriented approach. In 1977, at the behest of the Carter administration, the RANN program was canceled.

DARPA has continued to succeed as no other government research program has. Meanwhile, there has been a growing appreciation among many that the engineering and scientific problems of greatest importance to society are often interdisciplinary and exist on many levels. Research at the basic or foundational level, while often important, is also often far from sufficient (even when completely successful) to bring a discovery to the point where it will change society. In fact, just as it is possible to underemphasize basic research, which was the constant fear of 1970s-era NSF administrators, it is possible to overemphasize it. There is now a much greater awareness of the often-huge gaps between what industry is willing to do and what has been discovered in basic research, and there is a growing belief that science is made less relevant when these gaps are overlooked or deemphasized. By bridging these gaps, government scientists can transform society in ways that improve the lives of many.

Today, programs at the NSF and other agencies tasked with supporting or performing scientific research emphasize a more holistic approach toward research. Many government research agencies have created close ties between government, academic, and industry programs. Government funding often rewards researchers who have made these connections. This approach is still not as goal-oriented as that of DARPA, nor, on the whole, has this model succeeded to the extent that DARPA has, but it is a very different approach to science than the one championed by NSF administrators of the 1970s. While the idea behind RANN, which was conceived as a civilian version of DARPA, remains dormant, most government research programs are now more comprehensive in nature. These new programs are a sort of compromise between the short-lived, results-oriented approach of DARPA and RANN and the older, narrower emphasis on open-ended "basic" research that was pursued with such enthusiasm by the NSF.

The current model for conducting research is well illustrated by the federal biofuels programs. These programs are no longer described as research or even research and development (R&D) but rather research, development, and demonstration (RD&D). In these programs, basic research is pursued even as implementing technology is created. Next, or sometimes simultaneously with the R&D program, industry-government partnerships build and operate near-commercial sized projects both to demonstrate the commercial potential of the technology to interested parties and to identify potential problems with scaled-up processes and technologies. The RD&D approach to scientific research has been developed over decades and has been motivated in part by the sometimes-intractable problems associated with energy, including those associated with the development of biofuels. Cellulosic ethanol, biodiesel from algae, and biomass gasification are some of the technologies whose development continues to benefit from this more comprehensive understanding of scientific research. Not only has the development of biofuels benefited from this method of discovery, efforts to develop biofuels have served to inspire new ways of doing research.

Conclusion

Of all the different sources of energy that have been developed to produce electricity or power transportation, biofuels hold a special place. Their production requires the use of huge tracts of land and has the potential to introduce large changes in the price of food and feed. They are currently uneconomic when compared with fossil fuels, but financed by enormous government subsidies, programs to ensure their large-scale use are already under way. Increasingly, Americans are paying more to eat and paying more tax dollars to subsidize the production of ethanol, the production of which is contributing to rising food costs, rather than, say using those tax dollars to fund higher education or health care. The magnitudes of these changes are now a source of debate.

In addition to the diversion of agricultural resources to the production of fuel, large-scale biofuel production will also cause those responsible for managing forests, public and private, to reevaluate how best to use forestland. Intensive forestry management, now restricted to certain commercial tracts of land, may well be expanded to many of the nation's publicly held forests. Without these resources, the production of biofuels may remain too small to make a significant impact on the nation's fossil-fuel consumption.

Even these transformations will be insufficient without further and very substantial technological progress. Cellulosic ethanol, biomass gasification, and biodiesel from algae, the technologies that will fundamentally change the role of biofuels in the energy sector, remain under development. Progress is incremental, and it is unclear when these technologies will be ready for commercialization. Bringing them to market may require further substantial changes to tax law, agricultural practices, and even the genomes of some of the organisms used as feedstocks. These are complex problems, and most branches of science have found some application in the efforts to make biofuel use a reality.

The large-scale production of biofuels has generated heated debate about whether their manufacture will cause more harm than good. Biofuels promise a new way to produce energy, but whether the new way proves any better than the old remains to be seen.

Afterword:
An Interview with
Dr. Amani Elobeid on the
Economics of Ethanol

Dr. Amani Elobeid holds a doctoral degree in economics from Iowa State University. She is the international sugar and ethanol analyst with the Center for Agricultural and Rural Development (CARD) at Iowa State University. Her work centers on developing international sugar and ethanol models and conducting research on agricultural and trade policy. She has authored several papers on the impact of biofuels and multiple book chapters on food security, trade policy, and sugar markets.

Q: I've read your article, "Emerging Biofuels: Outlook of Effects on U.S. Grain, Oilseed, and Livestock Markets." I think it is interesting and informative, but before we talk about ethanol, could you talk a little about what an agricultural economist does?

Dr. Amani Elobeid
(Dr. Amani Elobeid)

A: Here at FAPRI [Food and Agricultural Policy Research Insti-tute] at Iowa State, we have economic models that cover agricultural commodities for the United States and also for the rest of the world. These are global models. We collect data on production, consump-tion, prices, and trade for each of these commodities for a number of major countries, and then we use the models to project out 10 years: what production will be, what consumption is projected to be, what prices are projected to be.

An agricultural economist, in general, will study and analyze the agricultural sector in different countries. What we also do here is to project out 10 years to what the agricultural economy will look like given certain macroeconomic conditions and certain policies.

Q: And the FAPRI model was developed at Iowa State?
A: Yes. Actually, FAPRI has collaboration with the University of Missouri at Columbia and some other universities—like [the University of] Arkansas and Texas A&M—and these models have been developed through the years with this collaboration. Iowa State University does the international side, and FAPRI Missouri concentrates on the U.S. agricultural economy.

Q: Ethanol is in the news a lot, both because of its contribution to the energy sector but also because of the impact its production has on food production. With respect to the ethanol market, would you agree that demand for ethanol is currently driven by its use as an oxygenate and its use as an octane enhancer? And about how much corn will it take to meet the current demand?
A: Part of the demand for ethanol is the fuel-additive part. That—I would have to check my numbers—is about 4 to 5 billion gallons of ethanol. When you incorporate the E10—that's 10 percent ethanol—that gets you into what we call "voluntary demand," because when you go to the pump you have a choice between fuel that doesn't contain ethanol and fuel that does. That will bring you up to about 14 to 15 billion gallons. We're looking at—depending on yields and other factors—we're looking at about 5 billion bushels of corn to produce this 14 to 15 billion gallons of ethanol. Once you start getting into E85—if there is a push here like they have in Brazil where you see more and more flex-fuel vehicles—then you get into the E85 market, and then our projections, as you've seen in the study, double that, so 30 billion gallons of ethanol.

Q: In producing this corn in order to meet the demand, how much of the corn is grown on new acreage, and how much is land shifted from other agricultural production such as soybeans? How does the production of this ethanol affect agricultural production in general?

A: What is happening now is that we're seeing a shift. You have, for example, in the Midwest with the high prices of corn and with the high demand for corn to be used in ethanol production, you're seeing land shifting out of other crops like soybeans and wheat and moving into corn production. So what's happening is corn is bidding away that land from soybeans and wheat and to a lesser extent other agricultural crops. For example, here in the Midwest, what is happening is that instead of having a corn-soybean crop rotation, they're doing corn-corn-soybean, or in some areas continuous corn. So you are using less land for soybeans. As you start talking about much larger volumes then you are going to start seeing land maybe coming out of CRP—

Q: CRP is the Conservation Reserve Program?
A: Yes, exactly. We may start seeing that. The new energy bill [the Energy Independence and Security Act of 2007] is talking about capping the corn-based ethanol at 15 billion and then looking at other sources for ethanol. They are looking at cellulosic. They're calling it "advanced biofuel," so it's cellulosic and some sugarcane-based ethanol, which would be imported from Brazil. So if this cellulosic ethanol comes from, let's say, a crop that has to be grown like switchgrass, then you will not only see switchgrass taking land away from other crops but expansion into "new land." But if this cellulosic comes from corn stover or wood chips, then you won't see that land usage shift as much, so the impact on land allocation will be a lot less.

Q: So what does it do to the price of other commodities as corn begins to take over the land and production of other commodities decreases? How does that affect consumers?
A: What you are seeing is an increase in the price of corn because of demand from the ethanol sector, and you are seeing more corn being grown and less soybeans, less wheat, and less of the other crops.

Their prices go up. If you look at the impact in isolation, just in the U.S., then you would calculate much higher soybean and wheat prices. But what is happening, and we have this in our model, is that other countries that are producing these commodities see those higher prices, and they respond. So, for example, yes, less soybeans are being grown in the United States because of the ethanol, but Argentina and Brazil are responding to those higher prices, and they are growing more soybeans. So the impact? You see an increase in prices, but the impact is dampened because other countries respond and fill the gap that the U.S. has left by reducing its production of soybeans.

Q: In general, food and energy prices are volatile. There is a large increase in the price of food right now, but there is some dispute about how much of the increase can be attributed to ethanol production and how much of the increase can be attributed to the increased cost of energy.
A: Right.

Q: Just how much of the current increases can be attributed to the production of ethanol?
A: It is very difficult to isolate or separate these. We know that ethanol production has increased the price of these agricultural crops, so naturally that trickles down to the consumer—not only in terms of the basic food side, but also in terms of the price of meat and livestock, because feed costs go up. But how much of that is solely due to ethanol, and how much is due to higher energy costs in general, that's hard to figure out. But what we've done in our study, "Emerging Biofuels," is to try by looking at that scenario—modeling ethanol and having it respond to high crude oil prices and looking at the predicted impact on food prices—and, in general, what we found is that, yes, food prices go up because of ethanol production, but not by the higher numbers that we've seen in real life, and that is be-

cause, most probably, the increasing energy prices have had a higher impact on food prices than ethanol. But we are only talking about 15 billion gallons. Once we start talking about 36 billion gallons of ethanol as we are seeing in the energy bill—even though a good portion of it is coming from advanced biofuels—that can have an impact on food prices, too. So as you see in our table 1 ("Emerging Biofuels"), we look at the impact and we see price increases hovering between 2 percent and 8 percent—I think the highest was eggs at 8 percent—but for all foods, from the baseline we had and the high crude oil scenario that we had, we found that the impact from ethanol was about 1.1 percent. So not large, but noticeable. I mean an 8 percent increase in egg prices is a noticeable increase.

Q: The price that farmers are paid for ethanol seems to be a healthy one. The market is growing so rapidly. But I'm not clear exactly how ethanol is priced. Does it just track the price of oil? So, for example, if gasoline is priced at three dollars per gallon, then because of ethanol's lower energy value, it would be sold for about two dollars per gallon, and if gasoline were to go to six dollars per gallon, then ethanol would be sold at four dollars per gallon. Exactly what is driving the price of ethanol?

A: What is driving the price of ethanol is demand and supply of ethanol. You have a demand for ethanol that increases the price of ethanol. Producers see that and so they produce more ethanol. When you start linking gasoline prices to ethanol, that would be the stage when you are using a lot more flex-fuel vehicles. When consumers go to the pump and they have a choice between no ethanol, 10 percent ethanol, and E85, at that point, when you have a volume large enough so that in a flex-fuel vehicle you see the difference between energy value, then you see the link between gasoline price and ethanol price become stronger. When you are looking at this range of production of up to 15 billion gallons, which is the fuel-additive market and the E10, that link is not that strong. So what is driving ethanol prices is, for example, in 2006, it was the additive market; it was the MTBE ban, and the rush by additive blenders to switch from MTBE to ethanol. And that really ramped up the demand for ethanol. Supply couldn't keep up. So ethanol prices shot up. At one time they were close to four dollars per gallon. But then as supply started to increase and started meeting demand, and in some cases exceeding demand, we saw ethanol prices go down. So it's just sort of market conditions, in terms of supply and demand, at this stage. Once we move into much larger volumes of ethanol, then you'll see that energy equivalent link between gasoline and ethanol prices.

Q: So with respect to the future ethanol market, how much do you think the ethanol market can grow? And what would be its limiting factors?

A: It just depends on what's going to happen in terms of policy and what's going to happen in terms of vehicle fleet composition. As long as we have a limited number of flex-fuel vehicles, then you're really looking at E10, i.e., 10 percent ethanol, and at a max, 20 percent, because there are some people who think that conventional cars will go up to E20. And so, if we're looking at—I have to, again, look up the numbers—but you have to see what fuel consumption will be in the next 10 years, and then what the expectation is in terms of E10 and E20, and that gives you a sense in terms of ethanol consumption, so it just depends. But who knows? Someone may come up with a hydrogen car or an electric car that will give you the range that you need in the United States for travel, and we won't be having this conversation. . .

Q: (*laughter*) Maybe. But in terms of increasing ethanol production . . . it's not like increasing oil production. Ethanol production takes a lot of land. Is the amount of land a concern in terms of going to 30 billion gallons? Previously, you said that some farmers are now planting corn continuously, one growing season after another, but surely that must take its toll on the land—

A: Very much so. And there are environmental consequences to doing that. Even ethanol production itself—it uses a lot of water so there are water concerns when it comes to producing ethanol—the environmental consequences of producing such large quantities of ethanol are significant. But again, if you look at it in terms of the United States, then those consequences, or negative impacts, are exaggerated. Once you look at large volumes of ethanol, then you are looking at other countries responding to that demand. Brazil, for example, if you have high demand for ethanol in the United States, that increases the price of ethanol. Brazil is right there waiting to export ethanol to the United States. The increase could be filled by importing ethanol from other countries. The new energy bill is looking at 15 billion gallons from corn-based ethanol and then the rest from other sources—cellulosic ethanol, imports from other countries, which is most probably Brazil.

Q: The Brazilians could export a lot of ethanol to the United States right now except for a tariff . . .
A: Exactly. But they have been exporting to the United States when ethanol prices have been really high, because even with the tariff, it's attractive for them to export ethanol to the U.S. Some of the ethanol from Brazil has been coming through the CBI [Caribbean Basin Initiative] countries, because they can export ethanol to the U.S. duty-free up to a certain limit, and some of that Brazilian ethanol has been coming through CBI countries.

Q: To return to the makeup of the future ethanol market, I notice in your paper that you don't hold out much hope for corn stover or switchgrass in terms of making a contribution to ethanol production. Could you explain why that is?
A: Well, the numbers don't add up. It costs a lot more to produce ethanol from corn stover or switchgrass, and what we've seen is that unless there is a subsidy, it doesn't make sense to produce it from these sources. Corn wins each time in competition with these

sources, given the numbers that we have in terms of how much it would cost to plant switchgrass. Right now, given the choice between switchgrass and corn, farmers will plant corn. Now if yields become much higher or if there is a change in technology where costs are reduced, then we could see a switch to other sources other than corn. Or if the government decides to subsidize these feedstocks higher relative to corn, then you could see that switch. But right now, the financial returns are just not there.

Q: So you see a booming ethanol market for at least a few years, and all of this demand will be met with food? Would that be a fair characterization?
A: You mean the demand for ethanol would be met with corn?

Q: Yes.
A: In the United States, yes. As long as corn is competitive in producing ethanol, then that is what will happen. Ethanol in the United States will be produced mostly from corn. But, for example, in Brazil they use sugarcane, and there are states in the United States that produce sugarcane, but it costs a lot more to produce sugarcane in the United States versus Brazil, so you're seeing maybe only one plant in the United States that is producing ethanol from sugarcane because it is so expensive here.

Q: OK. You see the corn production increasing to provide oxygenates, fuel additives, E10, and perhaps E20, and then the market would either level off, or there would be enough surplus ethanol to overcome the "E85 bottleneck." Could you talk a little about the E85 bottleneck?
A: The idea behind the E85 bottleneck is that you can produce more ethanol than E10, but if there is no market for it—for example, using flex-fuel vehicles—then where is that ethanol going to go? But at the same time, unless gas stations have E85 pumps so that it is more readily available, then people will not buy the flex-fuel vehicles that

use the E85. And so you have a two-sided situation: If there is a demand for E85, then you will start seeing E85, but on the other side, if there is no supply of E85, then there is no demand for E85. That is the problem, and that's the bottleneck that we talk about. Once you hit that 10 percent using conventional cars, then any more ethanol has to go into flex-fuel vehicles, and the bottleneck is that on the supply side, you have no E85 stations, and on the demand side you don't have enough flex-fuel vehicles to demand E85. Does that make sense?

Q: Yes, it does. It's a very complex situation determined partly by technology installed in automobiles, partly in energy sectors that have nothing to do with corn, and it's partly technology in the agricultural sector in terms of finding cost-competitive ways of producing ethanol using, say, corn stover, and it's a complex interplay between these factors, so it would be easy to make a wrong prediction. With that in mind, what do your models show in terms of ethanol production leveling off? What is your best guess as to what the future holds?
A: We are doing a new baseline that incorporates the new energy bill, so that would show different numbers than what our model shows now, which is basically leveling—our predictions are for 10 years—at a production level of 15 billion gallons of ethanol. Basically, E10 across the board, that is, 10 percent of fuel coming from ethanol. But now with this new energy bill, that story will change significantly. Since it calls for 36 billion gallons, then there will have to be drastic changes on the supply and demand side. Even if they produce 36 billion gallons, those 36 billion gallons will have to go somewhere or at least the 21 billion over and above the E10.

Q: OK. I noticed you didn't make a prediction.
A: (*laughter*) I don't know. It's safe to say that we will get to the 15 billion. Beyond that, it depends on what will happen with E85, cellulosic ethanol, and imports from Brazil. And the reason I hesitate

is that even though it is a mandate of 36 billion gallons, how it will be imposed is something I don't quite understand. There is no penalty, per se, just like when we had the 2005 energy bill there was a 7.5 billion gallon target by 2012, and we exceeded that, but if we hadn't what would have been the consequences? I don't know. So it's a mandate, but if we don't reach it, what will happen? I'm not clear on that.

Q: Thank you very much for sharing your insights.
A: You're welcome.

Chronology

1896 American inventor Henry Ford builds his first automobile. It is fueled with ethanol.

1900 A demonstration diesel engine is run on peanut oil at the Paris World's Fair.

In the United States, 41 percent of the workforce is employed in agriculture.

1908 Ford Motor Company begins production of the Model T. One of the most popular cars of all time, the Model T is capable of running on ethanol as well as gasoline.

1917–18 The production of ethanol fuel exceeds 50 million gallons per year.

1935 French ethanol production reaches 77 million gallons (290 million liters).

German ethanol production reaches 47 million gallons (180 million liters).

1937 Brazil ethanol production reaches 13.6 million gallons (51.5 million liters) per year.

1938 Approximately 2,000 service stations in the Midwest sell 10 percent gasoline-ethanol blends.

1945 Ethanol production begins a sharp decline worldwide.

1960 Fuel ethanol production down to almost zero in the United States.

1973   The EPA issues regulations aimed at phasing out the use of lead in gasoline, creating a new market for ethanol (as an octane enhancer). Most refineries, however, prefer MTBE.

1975   Brazil begins its National Alcohol Program (Programa Nacional do Ácool) to increase the use of ethanol as an automotive fuel. This marks the beginning of Brazil's modern biofuels market.

   Austrian researchers begin the first modern experiments into the use of vegetable oils as diesel-engine fuel.

1978   The Energy Tax Act seeks to create a market for ethanol by encouraging the use of gasohol (a 10 percent ethanol-gasoline blend) and a 40 cent per gallon ethanol subsidy.

1980   Fewer than 10 ethanol refineries are in operation in the United States.

1984   At the beginning of the year, there are 163 ethanol refineries in operation in the United States.

   The Surface Transportation Assistance Act of 1984 increases the ethanol subsidy to 50 cents per gallon.

   The Tax Reform Act increases the ethanol subsidy to 60 cents per gallon.

   By the end of the year, only 74 ethanol refineries remain in operation.

   The 50-MW Joseph C. McNeil Generating Station, then the world's largest wood-fired commercial power plant, begins operation in Burlington, Vermont.

1990   Congress passes the Clean Air Act Amendments (CAA) of 1990, requiring the use of gasoline additives—in practice this means MTBE or ethanol—to reduce air pollution.

1992   The Energy Policy Act of 1992 seeks to create ethanol markets by requiring operators of certain fleets of light vehicles to begin purchasing renewable fuels.

Establishment of the National SoyDiesel Development Board, later renamed the National Biodiesel Board, to support the creation of a biodiesel market in the United States.

1999   U.S. biodiesel production reaches 500,000 gallons (1.9 million liters).

2000   The EPA recommends that MTBE be phased out across the nation, thereby creating a large and dependable market for ethanol producers.

In the United States, 1.9 percent of the workforce is employed in agriculture.

2005   The Environmental Policy Act of 2005 mandates minimum amounts of renewable fuel in conventional gasoline, increasing from 4 billion gallons (15 billion liters) of ethanol in 2006 to 7.5 billion gallons (28 billion liters) by 2012.

2006   More than 80 percent of all new cars sold in Brazil are flex-fuel vehicles, designed to run on any ethanol-gasoline blend.

U.S. biodiesel production reaches 245 million gallons (927 million liters).

2007   The Energy Independence and Security Act calls for production of 36 billion gallons (136 billion liters) of ethanol per year in 2022.

The European Union proposes a binding target of 10 percent biofuels in the transportation sector by 2020.

CEOs of General Motors, Ford, and Daimler-Chrysler announce that 50 percent of all their new cars and light trucks will be flex-fuel vehicles by 2012.

2008   The European Union begins to reexamine its commitment to promoting biofuels in the transportation sector.

# List of Acronyms

DARPA	Defense Advanced Research Projects Agency
DDGS	dried distillers grains and solubles
EERE	Office of Energy Efficiency and Renewable Energy
EPA	Environmental Protection Agency
EU	European Union
FERC	Federal Energy Regulatory Commission
HHV	higher heating value
ISO	independent system operator
LHV	lower heating value
MTBE	methyl tertiary-butyl ether
NBB	National Biodiesel Board
NSF	National Science Foundation
RANN	Research Applied to National Needs
RD&D	Research, Development, and Demonstration
USDA	United States Department of Agriculture

 **Glossary**

air dry (weight)   the weight of a sample of biomass that contains 15 percent water when measured by weight

ash   the noncombustible solid component of a sample of fuel

base load   the minimum power load for an electrical system over a given time period

biofuel   any nonfossil fuel derived from biological sources

biomass   the living matter from which **biofuels** are produced

biotechnology   the application of biological science to the solution of practical problems

black liquor   a by-product of the process by which **pulp** is produced; black liquor serves as an important **biofuel** in the pulp industry

board feet   a unit of measure for lumber equaling a volume of wood that is 1 square foot by 1 inch thick

Btu (British thermal unit)   the unit of thermal energy sufficient to raise the temperature of one pound of water one degree Fahrenheit from an initial temperature of 39°F

C_3 plants   a class of plants that includes most temperate plants—these plants photosynthesize more slowly than **C_4 plants** under conditions of intense light or high temperatures

C_4 plants   a class of plants that photosynthesize more rapidly than **C_3 plants** under conditions of intense light or high temperatures

capacity   the ratio formed by the amount of power actually produced by an electrical generating station to the total amount of power that it could have produced if operated at full power during the time period of interest; also known as capacity factor

carbon sink   a "reservoir," either animate or inanimate, that removes carbon dioxide from the atmosphere

carbon source   a "reservoir," either animate or inanimate, that emits carbon, usually in the form of carbon dioxide or methane, to the atmosphere

cellulosic ethanol   the process by which ethanol is produced from cellulose and related materials that comprise the cell walls of grasses, trees, corn **stover**, and other **feedstocks**

cetane number   a measure of the auto-ignition properties of a diesel engine fuel

cofire   to burn a mixture of two different fuels, usually one **biofuel** and one fossil fuel

combined cycle   in electric-generating stations, a technology that uses the waste heat from a gas **turbine** to create steam to drive a steam turbine

dry mill   a mill for producing ethanol that produces distillers dried grains and solubles as its principal coproduct

feedstock   the biomass used to produce a **biofuel**

fuel treatment   the processing of **biomass** with the goal of reducing the risk of catastrophic forest fires

gasification   the conversion of solid fuels, in this case solid biofuels, to combustible gases

gasifier   a device for converting **biomass** to a combustible gas

greenhouse effect   the name of the phenomenon by which the emission of certain gases, most especially carbon dioxide and methane, increases the heat-retention properties of the atmosphere

greenhouse gases   those gases, principally carbon dioxide and methane, that contribute to global warming

green wood   unseasoned wood

higher heating value   a measure of the amount of thermal energy produced during combustion that takes into account the energy released as the water produced by the combustion process changes phase from vapor to liquid

landfill gas   the methane-rich gas produced as a result of the decomposition of the contents of a landfill

lignin   the material that binds together fibrous material of which woody plants are composed

lower heating value   a measure of the amount of thermal energy produced during combustion that does not take into account the energy released during the vapor-to-liquid phase change of the water produced by the combustion process

lubricity   the friction-reducing property of diesel engine fuel

MTBE (methyl tertiary-butyl ether)   a once-common fuel additive used to reduce engine knock and air pollution levels due to automobile emissions

net energy balance   the ratio formed by the amount of thermal energy released in burning a fuel to the amount of energy expended producing the fuel

oven dry (weight)   the weight of a sample of **biomass** after all water has been removed

peak power   the total electric power produced minus **base load** power production

petrodiesel   petroleum distillate used as fuel for diesel engines

poultry litter   poultry waste together with bedding material

pour point   the temperature at which diesel fuel ceases to flow

primary resource   **biofuels** that have been subjected to minimal processing

pulp   a fibrous material derived from wood or other material and used in the production of paper

secondary resources   **biofuels** such as ethanol and biodiesel that are derived directly from biomass feedstocks after significant processing

short ton   2,000 pounds

stover   that part of the corn plant that remains after corn has been harvested

switchgrass   a fast-growing hardy species of grass native to the North American prairies

tertiary resources   **biomass fuels** obtained from municipal or landfill wastes

turbine   a device used in a power plant that converts the linear motion of a **working fluid** into rotary motion

wet mill   a mill for producing ethanol that also produces corn oil, corn gluten meal, and corn gluten feed as coproducts

working fluid   in a steam power plant, the working fluid is used to drive the **turbine**

Further Resources

The production of biofuels touches on most branches of science and engineering, including physics, chemistry, genetics, agriculture, combustion, and economics. Issues of social justice and ethics are also important. The following sources offer different insights into a very complex field.

BOOKS

Avise, John C. *The Hope, Hype, and Reality of Genetic Engineering: Remarkable Stories from Agriculture, Industry, Medicine, and the Environment.* New York: Oxford University Press, 2004. One way to substantially increase biofuel production is to change the genetics of feedstocks in order to make them richer sources of energy. To some extent this has already been done, but it is possible to do much more. This book is about genetic engineering in general. It is not a book about biofuels in particular, but it does convey something of the potential contribution that genetic engineering can make to biofuel production.

Bungay, Henry R. *Energy, the Biomass Options.* New York: Wiley, 1981. After the oil crises of the 1970s, a large number of books were published describing various alternatives to the oil economy. As indicated by the title, this book describes the possible contributions of biomass, and as with all such books, reading this one illustrates how little has changed over the last few decades—it is valuable for providing a sense of perspective.

Hall, David O., Frank Rosillo-Calle, Robert H. Williams, and Jeremy Woods. "Biomass for Energy: Supply Prospects." In *Renewable Energy: Sources for Fuel and Electricity,* edited by Thomas Johansson and Laurie Burnham. Washington, D.C.: Island Press, 1993. "Biomass for Energy" provides a very revealing analysis of the relationships that exist between land use and biofuel production. More generally, this collection of research papers is a very rich source of information about renewable fuels. While some of the information is no longer current, much of the information in this book remains relevant and some is difficult to locate elsewhere.

National Academy of Engineering. *The Carbon Dioxide Dilemma: Promising Technologies and Policies.* Washington, D.C.: The National Academies Press, 2003. This book consists of a series of not especially technical papers by engineers, physicists, climatologists, economists, oceanographers, and others about global climate change and strategies to control it. It can also be read on the Web, albeit in a very nonuser-friendly format. URL: http://books.nap.edu/openbook.php?record_107988page=R10. Accessed on September 1, 2007.

Outlaw, Joe L., Keith J. Collins, and James A. Duffield, eds. *Agriculture as a Producer and Consumer of Energy.* Cambridge, Mass.: CABI Publishers, 2005. This collection of articles describes many aspects of how the agricultural sector consumes and produces energy. Some of the articles are very informative and contain information that is otherwise difficult to acquire.

Pahl, Greg. *Biodiesel: Growing a New Energy Economy.* White River Junction, Vt.: Chelsea Green Publishing Company, 2005. Although the book promotes rather than reports on the use of biodiesel, it is, nevertheless, filled with a number of useful facts and is worth reading.

Pool, Robert. *Beyond Engineering: How Society Shapes Technology.* New York: Oxford University Press, 1997. An interesting book that is concerned with the general problem of how society affects technology. While not about biofuels, it offers insight into the way that biofuel markets have (and have not) expanded.

Rosillo-Calle, Frank, Peter deGroot, Sarah L. Hemstock, and Jeremy Wood, eds. *The Biomass Assessment Handbook: Bioenergy for a Sustainable Environment*. Sterling, Va.: Earthscan, 2007. A somewhat technical introduction to the ideas involved in evaluating the potential of various biofuel feedstocks.

Tabak, John. *Coal and Oil*. New York: Facts On File, 2009. Biofuels are continually compared to fossil fuels, especially coal and oil, and are then classified as better or worse. To appreciate these arguments, it helps to learn more about these fossil fuels.

Williams, Michael. *Deforesting the Earth: From Prehistory to Global Crisis: An Abridgement*. Chicago: University of Chicago Press, 2006. This 543-page abridged version of the original work contains a good deal of hard-to-find information about the history of fire, forests, and wood fuel. It is a very ambitious work; it is not technical in nature, and it is highly recommended.

INTERNET RESOURCES

Many of the best biofuel analyses available online are published by the U.S. Department of Agriculture. Check their Web sites for articles in addition to the ones listed here.

Baker, Allen, and Steven Zahniser. "Ethanol Reshapes the Corn Market." In Amber Waves: The Economics of Food, Farming, Natural Resources, and Rural America, a publication of the U.S. Department of Agriculture. Available online. URL: http://www.ers.usda.gov/AmberWaves/April06/Features/Ethanol.htm. Accessed on December 28, 2007. The production of biofuels is changing the economics of corn production in ways that are complex and sometimes hard to predict. This is an analysis by two experts.

Coyle, William. "The Future of Biofuels: A Global Perspective." In Amber Waves: The Economics of Food, Farming, Natural

Resources, and Rural America, a publication of the U.S. Department of Agriculture. Available online. URL: http://ers.usda. gov?AmberWaves?November07/Features/Biofuels.htm#box. Accessed on December 28, 2007. Biofuel production is fast becoming a huge business with production distributed around the world. This is an excellent analysis of the topic.

Hayter, Sheila, Stephanie Turner, Kevin Comer, and Christian Demeter. "Biomass Cofiring in Coal-Fired Boilers." Available online. URL: http://www.nrel.gov/docs/fy04osti/33811.pdf. Accessed on December 28, 2007. Cofiring is currently the most common use of solid biofuels, and this somewhat technical primer is a good introduction. Everything but the appendices is accessible to the reader who has read this book.

IEA Bioenergy. "Biogas Upgrading and Utilisation." Available online. URL: http://www.iea-biogas.net/Dokumente/ Biogas%20final.pdf. Accessed on June 25, 2008. By converting biological wastes to biogas a very heterogeneous collection of feedstocks are converted into a very homogeneous fuel. This paper describes conversion technologies and possible applications of the resulting fuel.

Perlack, Robert D., Lynn L. Wright, Anthony F. Turhollow, Robin L. Graham, Bryce J. Stokes, and Donald C. Erbach. "Biomass as Feedstock for a Bioenergy and Bioproducts Industry: The Technical Feasibility of a Billion-Ton Annual Supply." Available online. URL: http://feedstockreview.ornl.gov/pdf/billion_ton_ vision.pdf. Accessed on December 28, 2007. This is one of the most frequently quoted and influential papers in this field. It is a technical document, but readers of *Biofuels* should be able to follow the discussion without difficulty.

Shurson, Jerry and Sally Noll. "Feed and Alternative Uses for DDGS." Available online. URL: http://www.ddgs.umn.edu/ar- ticles-industry/2005-Shurson-%20Energy%20from%20Ag%20C onf.pdf. Accessed on December 28, 2007. Despite the demand for ethanol and despite generous government production subsidies, the value of coproducts can mean the difference between profit and loss for ethanol producers. This narrowly focused article describes markets and potential markets for DDGS, the most common coproduct of ethanol production.

Tokgoz, Simla, Amani Elobeid, et al. "Emerging Biofuels: Outlook of Effects on U.S. Grain, Oilseed, and Livestock Markets." Available online. URL: http://www.card.iastate.edu/publications/DBS/PDFFiles/07sr101.pdf. Accessed on April 15, 2008. This is the paper to which Dr. Elobeid refers in her interview. It is important research into many important issues associated with the rapidly growing biofuel markets and their effects on the broader economy.

Valdes, Constanza. "Ethanol Demand Driving the Expansion of Brazil's Sugar Industry." Available online. URL: http://www.ers.usda.gov/Briefing/Sugar/sugarpdf/EthanolDemandSSS249.pdf. Accessed on December 28, 2007. A brief and highly informative primer on ethanol and sugar production in Brazil, the pioneer and still a leader in biofuel production.

Wescott, Paul C. "U.S. Ethanol Expansion Driving Changes Throughout the Agricultural Sector." In Amber Waves: The Economics of Food, Farming, Natural Resources, and Rural America, a publication of the U.S. Department of Agriculture. Available online. URL: http://www.ers.usda.gov/AmberWaves/September07/Features/Ethanol.htm. Accessed on December 28, 2007. An important and highly readable analysis of some of the many implications of the U.S. ethanol boom.

Index